Life History Invariants

Oxford Series in Ecology and Evolution
Edited by Robert M. May and Paul H. Harvey

Life History Invariants
Some Explorations of Symmetry in Evolutionary Ecology

ERIC L. CHARNOV

Department of Biology
University of Utah
Salt Lake City

Oxford   New York   Tokyo
OXFORD UNIVERSITY PRESS
1993

Oxford University Press, Walton Street, Oxford OX2 6DP

Oxford   New York   Toronto
Delhi   Bombay   Calcutta   Madras   Karachi
Kuala Lumpur   Singapore   Hong Kong   Tokyo
Nairobi   Dar es Salaam   Cape Town
Melbourne   Auckland   Madrid
and associated companies in
Berlin   Ibadan

Oxford is a trade mark of Oxford University Press

Published in the United States
by Oxford University Press Inc., New York

A catalogue record for this book is available from the British Library

Library of Congress Cataloging-in-Publication Data
Charnov, Eric L., 1947–
Life history invariants : some explorations of symmetry in
evolutionary ecology / Eric L. Charnov.
(Oxford series in ecology and evolution)
Includes bibliographical references (p.147) and index.
1. Evolution (Biology)   2. Symmetry (Biology)   3. Sex differences.
I. Title.   II. Series.
QH371.C485   1993   575 — dc20   93–7170
ISBN 0–19–854072–8 (h/b)
ISBN 0–19–854071–X (p/b)

Typeset by Apex Products, Singapore
Printed in Great Britain by
Bookcraft Ltd, Midsomer Norton, Avon

This book is dedicated to George C. Williams, founder of modern life history theory, who admonished us to use natural selection very carefully . . . but none the less to use it at every turn, and to R. J. H. Beverton, who pioneered the use of dimensionless numbers for the comparative study of life histories.

Preface

This book explores several roles that principles of symmetry play, or can play, in evolutionary ecology. These symmetry principles can be particularly valuable to our understanding of the structure of life histories. In the most general sense, symmetry means that one or more attributes of an object of interest remain unchanged under specified transformations, i.e. the character is invariant or illustrates a conservation principle. Invariants and symmetry play central roles in physical science (Pais 1986; Stewart and Golubitsky 1992), but have been much less often used by population or evolutionary biologists more concerned with differences between rather than similarities among species (a clear exception is D'Arcy Thompson's classic book, *On growth and form*). An evolutionist may say, for example, that an enzyme is conserved—meaning that the enzyme is approximately the same molecule when viewed over a specific evolutionary time frame. However, this book does not examine these kinds of conserved characters, but characters which are defined by the relationship of various elements of life histories.

Population biologists are familiar with many transformations such as changes in age, body size (within a species or phylogeny), sex (substitute male for female), population size (total or density), geography, and time itself. Questions about symmetry would ask which attributes are unchanged under one or more of these or other specified transformations. For example, Mendelism refers to an inheritance symmetry where the heterozygote (A/a) gives equal numbers of each gamete type. An example of temporal symmetry is a stable age distribution, where the proportions of the age groups are unchanged (invariant) over time. In addition, within specified taxonomic boundaries many physiological and life history variables scale with adult body weight W^P; if the exponent P is the same for two variables, their ratio is necessarily invariant with body size ($W^P W^{-P} \propto W^0$). Finally, demographic variables like age at maturity or the adult mortality rate may sometimes be independent of (invariant with respect to changes in) population size, demonstrating density independence in the parlance of population biologists.

This symmetry theme is used in several distinct ways in the chapters. First, though, we must make some general assertions.

(1) Invariance in the relationship between pieces of a life history will necessarily be only in a statistical sense; most, if not all, life history variables estimated for field populations are noisy.

(2) The most natural way to express the relationship between elements of a life history is in terms of dimensionless numbers, an example being weaning weight divided by adult weight, a number equal to approximately 1/3 for a wide variety of mammals.

(3) Sometimes it is useful to develop symmetry ideas, even if the invariance is not quite true; afterwards we can always ask what additional factors might be breaking the symmetry.

(4) Symmetry in a causal process may produce asymmetry in an outcome. For example, the 0.75 body size allometry of basal metabolic rate has often been viewed as the outcome of a deeper-level symmetry about something else and body size.

(5) We can search for invariant outcomes in the relative structure of life histories, and *then use evolutionary life history theory to probe for the underlying symmetries which we guess to be causal*. Sometimes, as with sex ratio, we can begin with a deeper-level symmetry and then work in the other direction.

(6) Asymmetry in a large way can often be the most useful insight. Anisogamy generating sexual selection and age structure generating senescence are two major examples.

The problems and concepts are introduced in Chapter 1 with several real world and hypothetical examples. In Chapter 2 a well-understood inheritance symmetry (everyone has one mother and one father) is used to investigate evolution of sex ratio and related problems. Sex ratio is particularly illustrative since non-equal sex ratios often reflect the basic inheritance symmetry (giving 1:1) combined with asymmetry in other causal factors. In Chapter 3 we illustrate how the temporal age distribution symmetry inherent in a non-growing (stationary) population greatly simplifies the study of alternative male life histories. In Chapter 4 we focus on fish, shrimp, reptiles, and other animals that continue growing after maturation, and review some general patterns which interrelate their growth, maturation, and mortality. Female mammals are discussed in Chapter 5 where we look at body size allometry and the resulting body size invariants which follow from shared scaling exponents. Sexual dimorphism (an asymmetry) in adult body size is also considered. Chapter 6 is a brief look at what some particular population size and body size invariants might tell us about population dynamic processes. Chapter 7 is about a near-universal asymmetry, that of age itself where older is worth less, in a fitness sense, than younger. Ageing, or senescence, is the evolutionary end result. In this chapter we use results from the previous six chapters to point toward biological systems which seem particularly useful for ageing studies. Chapter 8 summarizes and points to the future.

Finally, what does this book not do? It is heavily biased towards animals compared with plants. It does not cover all, or even most, of life history evolution and ignores many important topics about which I have nothing particularly interesting to say. The evolutionary models developed here are all phenotypic, and no formal population genetics is included. Except in Chapter 6, normalizing natural selection in stationary populations is assumed, so that the modelling considers neither population nor evolutionary *dynamics*; only *statics* are considered. What remains is a series of explorations where life histories, ecology, physiology, and natural selection meet around the common theme of invariance or symmetry. I am aware that notions of symmetry go much deeper than anything attempted in this book; it should be considered an invitation to the more mathematically adept. This book uses mathematics only at the level of calculus (but for good reasons) and assumes familiarity with the basics of population and evolutionary ecology, particularly life history and evolutionary game theory (Charnov 1982; Maynard Smith 1982; Roff 1992; Stearns 1992; Charlesworth 1994).

The chapters are not sequential. Indeed, I suggest reading Chapter 1 to start, followed by the three groupings, in any order, of Chapters 2 and 3, Chapter 4, and Chapters 5 and 6. Chapters 7 and 8 should be read last. The chapters are not completely independent, however, so that the reader should be prepared to 'go to Fig. *x* on p. *y*' in another chapter from time to time.

Summit Park, Utah   E.L.C.
January 1993

Acknowledgements

For reading all (or part) of the manuscript and providing useful comments I thank David Berrigan, James Bull, Vince Eckhart, James Ehleringer, Charles Fowler, Charles Godfray, Mart Gross, Paul Harvey, Robert May, Linda Partridge, Amy Roeder, Alan Rogers, Cindy Sagers, and Stephen Stearns. For discussion of some of the issues raised herein I thank R. J. H. Beverton, Kristen Hawkes, Derek Roff, Richard Shine, John Maynard Smith, and George C. Williams. Special thanks are due to David Berrigan, good friend and colleague, who helped the book in innumerable ways and helped me to think about these issues for almost 4 years (1988–1992). The staff at OUP always went the extra mile, as did Jeanette Stubbe, who mastered my scribble and cheerfully typed the many drafts. Julia Riley turned it into English and removed approximately 80 per cent of my original semicolons. Kerry Matz drew the diagrams that the computer choked on. The Biology Department at the University of Utah paid some of the bills for typing, diagrams, and duplication. Finally, I thank my writer wife Julia Laylander who understands the demons, and of course Agatha and Dasher, who expect the royalties to be used exclusively to keep them in bones.

Contents

1

Introduction: invariants imply deeper symmetries

1.1 Introduction (with examples)

This book is about invariants, characteristics that do not change, in the life histories of plants and animals. When (population) biologists think of invariance or symmetry, they often think of geometric symmetry such as right–left or radial, the symmetry of repeating units such as coiling in snails or the modular construction of many plants and sessile animals, or the symmetry in gaits of running animals (Stewart and Golubitsky 1992). But symmetry is a much more general concept, meaning that some attributes of an object of interest remain unchanged under specified transformations. Symmetry can have deep mathematical and physical underpinnings, and has proved to be an immensely fruitful approach in physical science (see Pais (1986) for a popular overview of particles and forces).

In this book we shall use invariance and symmetry in an informal way, asking first *which* attributes of life histories are unchanged under *what* transformations, second *why*, and third *what* can the invariance tell us about other features of the life history? Only a subset of these questions will be under discussion at any point in the book. I tend to use the terms invariance and symmetry more or less interchangeably. Something will be called invariant (or an invariant) if it does not change under the specified transformation(s). The invariance may well have implications for other biological processes (e.g. population dynamics). However, the invariance itself will be caused by some deeper features of the life history (the *why*?). These deeper features will, I claim, often be of such a form that only certain kinds of trade-off relations underlie the evolutionary structure of the life history. The allowed trade-off relations will themselves show symmetries, i.e. they will hold some feature(s) constant under their transformations. Thus symmetry is proposed as a (strong) guide to understanding some features of life histories.

For three concrete examples of invariance consider Figs. 1.1 (fish), 1.2 (birds), and 1.3 (mammals). Figure 1.1 is for a sex-changing fish studied at two localities; the species is protogynous, i.e. an individual changes from female to male as it becomes larger (older). The two localities have very different body size distributions for the adults; indeed, on average, the adults in the upper panel are about three times the weight of those in the lower. The males in the lower panel have about the average body size of the females from the upper panel, whereas all the males from the upper panel are larger than any fish from the lower panel. However, the breeding sex ratio is fixed at approximately 25 per cent male for each of these two localities. The sex ratio is invariant under the transformation of average adult body size for this species. In Chapter 2 it will be shown (eqn. (2.5) or eqn. (2.8)) how this approximate invariance follows from a basic inheritance symmetry (everyone has one mother and one father).

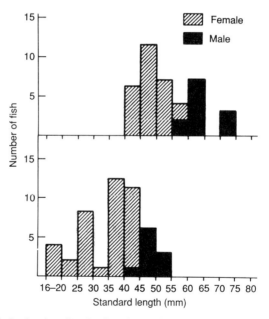

Fig. 1.1 Adult body size distribution for male and female *Anthias squamipinnis* from two locations (800 m apart) on the same Aldabra Island reef. The species is a protogynous sex changer. The largest 20–30 per cent are males, but the body sizes differ dramatically between the two locations. Thus the breeding sex ratio is invariant to changes in the population body size distribution. Data from Shapiro (1979).

Table 1.1 summarizes data on the average adult instantaneous mortality rate M (yearly survival is $e^{-M(1\,\text{year})}$) and yearly clutch size b in daughters for a variety of birds representing diverse ecologies and body sizes (sparrow versus albatross). Both M and b have dimensions of time^{-1} and so their ratio is dimensionless. Figure 1.2 shows a log–log plot of b versus M; it is almost perfectly linear with slope not different from unity, indicating that the ratio of b to M is a constant—an invariant—among these species. While it is harder to label just what transformation is indicated here, a relationship among elements of a life history (i.e. b/M) is invariant to the evolutionary changes which resulted in the albatross and sparrow we see today: both b and M change greatly, but their ratio does not. Further, we can use the invariance in the ratio to argue for an additional invariance. b/M is the average number of daughters produced over a female's adult life-span, provided that b and M are approximately independent of age. In a population of unchanging size, at least on average, the female must have exactly one daughter survive to adulthood. Let $S(\alpha)$ be the chance that a daughter survives to age α, the age at maturity or first reproduction. By the stable population assumption $S(\alpha) = M/b$; for all these species $S(\alpha) \approx 0.2$, even though they differ greatly in age at maturity (Ricklefs 1969, 1977).

Figure 1.3 shows a log–log plot of yearly clutch size b (in daughters) versus the age at maturity α, from two studies on mammals (Ross (1992)

Table 1.1. Yearly clutch size b (in daughters) and adult instantaneous mortality M for representative species of birds.

Species	Locality	M (years^{-1})	b (years^{-1})
Wandering albatross	S. Georgia Island	0.044	0.17
Gannet	England	0.063	0.40
Black-and-white manakin	Trinidad	0.12	0.50
Kittiwake	England	0.13	0.75
Yellow-eyed penguin	New Zealand	0.17	0.81
Shag	England	0.22	1.00
Brown pelican	Southeastern USA	0.22	0.43
Scrub jay	Florida	0.22	0.57
Great blue heron	Eastern USA	0.25	1.00
Red-shouldered hawk	Eastern USA	0.37	0.94
Barn swallow	Eastern USA	0.56	2.99
Black-capped chickadee	Eastern USA	0.69	3.00
Robin	Eastern USA	0.69	2.59
Tree sparrow	Poland	0.80	6.60

From Ricklefs (1977); see for original sources.

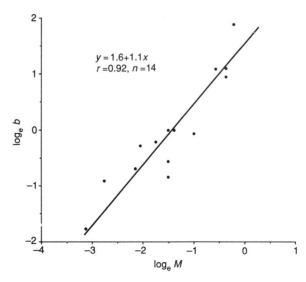

Fig. 1.2 Functional regression (major axis) of yearly clutch size b on yearly adult instantaneous mortality rate M for 14 bird species. Both b and M have dimensions of time $^{-1}$ and so their ratio is a pure number (i.e. dimensionless). A slope not different from unity means that b/M is invariant (approximately 5) for these species. The data and species are summarized in Table 1.1. See text for further discussion.

for primates; Hennemann (1983) for non-primates). The fitted relation has a slope of 1, indicating that the dimensionless number αb is approximately invariant. While it is harder to give an intuitive meaning to this number (compared with, say, b/M), the underlying transformation is adult body size between species: large primates have the same αb value as small ones, and the same value as shown by many even more distantly related mammals of various sizes. The αb number will be discussed in more detail in Chapters 5 and 6 (e.g. eqn. (5.9)).

These three examples using fish, birds, and mammals illustrate five important points. First, the invariance will often be with respect to the relationships between elements of the life history, leading us to form dimensionless numbers made up from the usual life history variables. M and b both have units of time $^{-1}$, and so their ratio is a pure number whose value is independent of the measurement units for b and M (Charnov and Berrigan 1991a,b; Stephens and Dunbar 1993). Second, b/M is invariant under a transformation which is not particularly well defined, while the body size transformations of Figs. 1.1 and 1.3 are fairly well defined. We have much to learn about what constitutes interesting/useful transformations. Third, we have not yet given

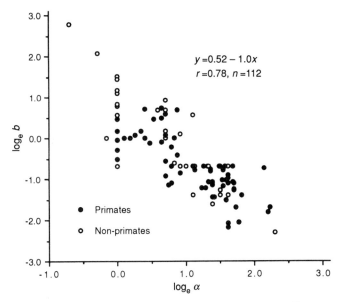

Fig. 1.3 Yearly clutch size b (in daughters) versus the age α at first reproduction for mammals, separated into primates (72 species) and non-primates (40 species). α has units of time; b has units of time^{-1}. Under the reasonable assumption that α is measured with much less error than b, regular regression shows a slope of -1 indicating that the pure number αb is (approximately) invariant, and has the same value (approximately 1.71; 95 per cent confidence interval 1.4–2.0) for primates and non-primates. Analysis of covariance shows no difference in either slope or intercept between the two groups. The transformation is adult body size between species. Data from Ross (1992) for primates and from Hennemann (1983) for non-primates.

any causal reason why the invariance exists; it may often prove easier to find invariants than to understand what generates them. However, the question posed is not to understand what causes M or b to take particular values, but to understand why their ratio does not change under the sparrow–albatross transformation. Fourth, we can often combine causal factors to find new invariants: if b/M is a constant, then in a non-growing population so is $S(\alpha)$. The causal factor is the stable population (perhaps). Fifth, the invariance holds only in a statistical or probabilistic sense. Most life history variables estimated for field populations are noisy (Fig. 1.3 shows a level of precision common for field data, $r \approx 0.8$). How constant is constant enough to be considered invariant is worthy of much thought, considering that noisy field parameters are our stock in trade and will be forever.

In contrast, some of the most interesting patterns in evolutionary ecology follow from the existence of *well-defined asymmetries*, the two most conspicuous being anisogamy generating sexual selection (Orians 1969; Parker *et al*. 1972; Trivers 1972) and age structure generating senescence (Medawar 1952; Williams 1957; Rose 1991). In applying symmetry arguments it is well to keep in mind that both symmetric and asymmetric (symmetry breaking) causal factors may interact to produce the outcome; this is particularly true for sex ratio evolution.

In the rest of this chapter we focus on the kinds of invariants that we might look for, the kinds of transformations that might be involved, the possible implications of the invariance for other processes, and the ways that we might begin to investigate and understand causal relations. It will almost always be true that invariance at the level of the relationships between elements of life history (e.g. b/M) implies symmetry at a deeper level of causal factors.

1.2 Population dynamics

Two transformations characterize populations: size (total or density) and time. Under change in size we imagine that some demographic variables such as birth or death rates alter in magnitude with population size (density dependent) while others do not (density independent). A symmetry approach looks for rules about *what* falls into each class. Population biologists have long looked for such symmetries (asymmetries) (e.g. Fowler 1981, 1987; Sinclair 1989).

Basic ecology courses usually consider one temporal invariant, the stable age distribution (SAD). Given a fixed set of age-specific (stage-specific) birth and death rates, a population followed through time usually approaches a state where the relative numbers of each age (stage) category remain fixed. Computer exercises convince our students of this result, but we are quick to follow with a disclaimer that the SAD condition is rarely realized in nature, except perhaps as an average over several generations, and is to be treated as an idealized, if useful, fiction. For example, an intrinsic rate of increase for specified demographic variables, independent of detailed specifications of actual age distributions, can only be calculated for an SAD.

A population unchanging in size with an SAD is said to be stationary. If $L(x)$ is the probability an individual is alive at age x ($L(0) = 1$), then Fig. 1.4 illustrates an interesting equivalence relation. We plot the $L(x)$ schedule and on the same graph place the number of individuals in each age category. We now divide the number of individuals in each by the number of age zeros, the first age category, so as to normalize the

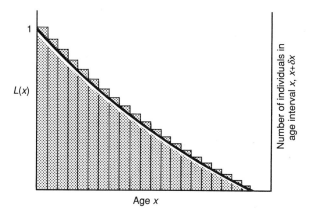

Fig. 1.4 A stationary population has a stable age distribution and is unchanging in size. For such a population the $L(x)$ schedule (—) is the same as the standing age distribution (shaded columns); this *equivalence* is very useful in the derivation of ESS results as shown in Chapters 2 and 3.

data (i.e. number of age zeros = 1, etc.). In a stationary population these two relations will be exactly the same, i.e. the $L(x)$ curve will fall on top of the normalized number of individuals in each age category. This equivalence relation, itself a symmetry, is very useful as a theoretical tool as illustrated below and used in Chapters 2 and 3. The reflection of mortality rates by the standing age distribution has long been recognized as an approach to estimating mortality rates (e.g. Seber 1973).

1.3   Fitness and life history evolution

In this book stationary populations are assumed (except in Chapter 6); stabilizing natural selection is also assumed. In this way I simply ignore the more complex problems of population or evolutionary *dynamics* and focus on more tractable problems of comparative *statics*. Under a wide variety of autosomal genetic systems (single locus, polygenic, etc.) the evolutionary problem then reduces to the maximization of the net reproductive rate R_0 (or the two-sex equivalent) in the face of constraints or trade-offs among its component parts (e.g. Lande 1982; Charnov 1986; Stearns 1992; Charlesworth 1994). One necessary constraint is that $R_0 = 1$ for a non-growing population. This means that one or more demographic variables must show density dependence. By using R_0 as both a fitness measure and a population dynamic constraint ($R_0 = 1$ in equilibrium), we are forced to deal explicitly with the important issue of *where* in the life history density dependence is present. I usually implement this population dynamic constraint by putting all density dependence into survival of the immatures, an assumption developed at several places in the book.

R_0 can be written as:

$$R_0 = \int_{\alpha}^{\infty} L(x) b(x) \, dx \qquad (1.1)$$

where α is age at maturity, $L(x)$ is probability of survival to age x, and $b(x)$ is birth rate (in daughters) at age x $(x \geqslant \alpha)$. R_0 can be rewritten as

$$R_0 = S(\alpha) V(\alpha) \qquad (1.2a)$$

or

$$\log_e R_0 = \log_e V(\alpha) + \log_e S(\alpha) \qquad (1.2b)$$

where $S(\alpha)$ is the chance of living to age α and $V(\alpha)$ is the average number of daughters produced over a female's adult life-span if she is alive at age α.

$V(\alpha)$ should increase with α, while $S(\alpha)$ must decrease. The optimal (or Evolutionarily Stable Strategy, an ESS) age at maturity $\hat{\alpha}$ (Bell 1980; Stearns and Crandall 1981; Maynard Smith 1982; Roff 1984, 1986, 1992; Charnov 1990; Kozlowski 1992; reviewed by Stearns 1992) is where $\partial \log_e R_0 / \partial \alpha = 0$ (or $\alpha = 0$). This yields the rule

$$\frac{\partial \log_e V(\alpha)}{\partial \alpha} = - \frac{\partial \log_e S}{\partial \alpha} .$$

Let $S(\alpha) = \exp[-\phi(\alpha)]$ so that $\log_e S(\alpha) = -\phi(\alpha)$; then we have in ESS:

$$\frac{\partial \log_e V(\alpha)}{\partial \alpha} = \frac{\partial \phi(\alpha)}{\partial \alpha} . \qquad (1.3)$$

This solution is illustrated in Fig. 1.5; it should be noted that density dependence, which falls on the very young, will only affect the ϕ function for small α. This will not alter $\hat{\alpha}$, provided that the slope of the ϕ function is not altered for larger α. Density dependence on mortality for

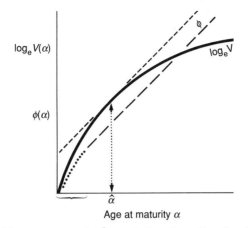

Fig. 1.5 The ESS age at maturity $\hat{\alpha}$ maximizes $\log_e R_0$, the difference between $\log_e V(\alpha)$ and $\phi(\alpha)$, as developed in the text (argument leading to eqn. (1.3)). This is where the slope of the ϕ function (——) equals the slope of the $\log_e V$ function (—). If density dependence affects mostly survival of the very young, it will alter ϕ for small α only (bracketed area) and probably will not change the slope of the ϕ function for larger α; thus $\hat{\alpha}$ will be density independent.

the very young only causes the ESS α to become density independent, i.e. invariant with respect to population size. Charnov (1990) discusses how density dependence on the $V(\alpha)$ function may often also not affect $\hat{\alpha}$, particularly if the adult instantaneous mortality rate M is itself density independent.

Therefore I use R_0 as a fitness measure (i.e. maximize R_0 in the face of trade-offs) and a population dynamic constraint ($R_0 = 1$). We confront head-on the issue of where in the life history to put density dependence; I choose survival of the very young as a plausible and defendable general assumption (Fowler 1981, 1987; Sinclair 1989), but I also recognize that other assumptions may be appropriate in particular settings. Many workers have noted that the $R_0 = 1$ assumption alone allows predictions about life histories, particularly if numerical values are known for all of the other components of R_0 (Sutherland *et al.* 1986). Indeed, this has often been used to predict immature mortality rates (see Lack (1954) for birds). Murray has used the $R_0 = 1$ assumption to predict clutch size in birds with reasonable success (Murray and Nolan 1989; Murray *et al.* 1989). He appears to believe that the prediction supports an evolutionary model for clutch size when in fact it simply reflects the demographic constraint of $R_0 = 1$. Its general success is non-trivial, however, for it supports the use of R_0 as a fitness measure; $R_0 \approx 1$ is necessary for the use of this parameter as a fitness measure.

1.4 Inheritance and sex allocation

Mendelian segregation under diploidy is the most useful inheritance symmetry; combined with Hardy–Weinberg equilibrium and fitness it yields population genetics. Mendelism for autosomes plus two sexes yields another useful symmetry. This is the fact, first noted by Fisher (1930), that half the genes passed to zygotes in any generation come from males and half from females (everyone has one mother and one father). This creates frequency-dependent natural selection on the sex ratio, as large reproductive gains accrue to the minority sex (or the mothers producing them). This one mother–one father symmetry is explored in Chapter 2; here I merely indicate the kinds of predictions that follow from it, particularly when it is combined with the $L(x)$-age distribution symmetry discussed earlier for a stationary population (Fig. 1.4).

Consider the case of sex reversal, particularly protogyny in which an individual reproduces first as a female and then changes sex to reproduce for the rest of its life as a male. Upon closer examination, many protogynous fish species have been found to have populations that consist of sex changers and pure males (Choat and Robertson 1975; Robertson and Warner 1978; Warner and Robertson 1978). The typical life history is illustrated in Fig. 1.6. The young fish (initial phase (IP)) may be either male or female; the older ones (terminal phase (TP)) are male. The IP fish may often have different coloration from the TP fish; IP males usually use different reproductive tactics from TP males. The ESS proportion q going down the pure sex (male) pathway equalizes

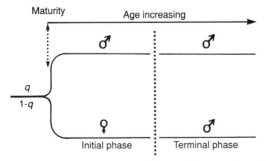

Fig. 1.6 Many labroid fish have populations that consist of a mixture of protogynous sex changers and pure males. The fish often come in two color forms, initial phase (IP) and terminal phase (TP). IP fish may be male or female, while all TP fish are male. q is the proportion of the IP fish who are male.

the reproductive gains over the two pathways. In a stationary population this is equivalent to (Charnov 1989*b*, 1982, p. 174)

$$q = \frac{H}{1 + H}$$

where H is the proportion of the females who mate with IP males in a single breeding season. Therefore, stationarity means that we need not measure life-time reproduction for an individual, but can substitute a prediction about what the females, as a group, do at a single point in time. One other prediction for this system (developed by Charnov (1989*b*)) is that the sex ratio, i.e. the proportion of males among the breeders, will be less than 1/2 under the assumption of a stationary population. The proportion can approach 1/2 from below but theoretically cannot penetrate 1/2. Figure 1.7 illustrates this for 18 species of labroid fish from Australian and Caribbean coral reefs. The data strongly support the theory: about half the species fall in the range 40–50 per cent males, but none fall in the range 50–60 per cent males. Fifty per cent males is indeed a barrier from the lower side. I use this admittedly exotic sex ratio example to illustrate the usefulness of the one mother–one father symmetry combined with a stationary population (Fig. 1.4). These tools will be used extensively in Chapters 2 and 3.

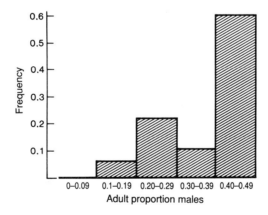

Fig. 1.7 Frequency diagram of the breeding sex ratio for 18 species of labroid fish, each of which has a population consisting of a mixture of protogynous sex changers and pure males as illustrated in Fig. 1.6. Theory predicts that the adult proportion of males should be less than 0.5; the sex ratio is clearly biased as predicted. Data from Robertson and Choat (1973), Choat and Robertson (1975) Robertson and Warner (1978), and Warner and Robertson (1978).

1.5 Relative timing (and body size) variables

Many life history variables have units of time or time $^{-1}$. Examples for time are age at weaning, age at maturity α, and maximum (or average) life-span. Examples for time $^{-1}$ include instantaneous mortality rates M (for adults), yearly clutch size b, relative growth rates (the Bertalanffy coefficient k), and the intrinsic rate of population increase. Some other life history variables have units of length or weight; examples include size at birth, size at weaning (at independence from the mother), size at maturity, and asymptotic size. Dimensionless ratios or products within the time or size groups can be constructed, and we can ask if the values are invariant over some transformations. Then we can use evolutionary life history theory to ask: why? Chapters 4–6 are devoted to these variables; here I introduce the problem using one such number.

Let α be the age at first reproduction, measured from independence from the mother (e.g. weaning or hatching), and let M be the average adult instantaneous mortality rate (Beverton 1963). In Fig. 1.8 $1/M$ is plotted versus α for a large sample of bird and mammal species showing the best fit regressions through the origin (proportionality). Log-log plots give a slope of 0.98 for the mammals, which is not different from proportionality (a slope of unity). The birds have a slope of 1.2. This is statistically different from unity, indicating that birds with large α have relatively larger average adult life-spans $1/M$. Proportionality would mean that

$$\frac{1/M}{\alpha} = (\alpha M)^{-1}$$

is a constant, i.e. an invariant within the group. According to the plots in Fig. 1.8 this is true for mammals but only approximately true for birds. Of much greater interest is that, for a fixed α, birds have average adult life-spans about double that of mammals (2.47 versus 1.42). The $(\alpha M)^{-1}$ values for birds and mammals also differ greatly from those for other vertebrate groups, as illustrated in Fig. 1.9 (data from Charnov and Berrigan (1990) and from Chapters 4 and 5 of this book). Lizards and snakes have $(\alpha M)^{-1} \approx 0.8$ (Shine and Charnov 1992), while fish (and pandalid shrimp (Charnov (1979c)) have $(\alpha M)^{-1} \approx 0.5$. It should be noted that, for a given age at maturity, birds have adult life-spans about twice those of mammals, three times those of snakes and lizards, and five times those of fish and shrimp. Figure 1.9 says nothing about the actual values of α or M; indeed, there are fish with ages at maturation

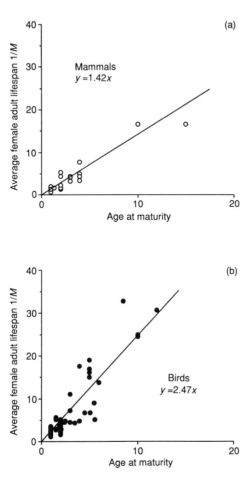

Fig. 1.8 Average female adult life-span $1/M$ plotted against age at maturity α for (a) 26 mammal species (data from Millar and Zammuto (1983) and Promislow and Harvey (1990)), listed in Fig. 5.5, and (b) 66 bird species (data from many sources). Equations are the best-fit regressions through the origin with 95 per cent confidence intervals as follows: birds, 2.26–2.68; mammals, 1.30–1.54. The mammals give the same slope even if we exclude the two high points. Log–log plots show a slope not different from unity (proportionality) for mammals but a slightly higher slope (1.2) for birds. See text for further discussion.

of 10–20 years (Beverton & Holt 1959; Beverton 1963), the same magnitude as elephants and well beyond that of most birds. It is the relation between α and M which differs between groups: upon reaching the maturation age of 10 years, a bird has an average of about 25 years to live, whereas a fish has about 5 years.

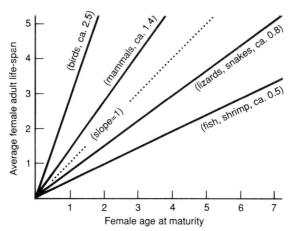

Fig. 1.9 Average adult life-span $1/M$ versus α for all groups discussed in this book (mammals (Chapter 5), and fish, reptiles, and shrimp (Chapter 4)). For contrast, birds are treated as having a proportional relation (see text for qualifications). Lizards and snakes are pooled. Shrimp data are for both sexes. The number on each line refers to the estimated slope, i.e. $(\alpha M)^{-1}$. After Charnov and Berrigan 1990.

The variation in $(\alpha M)^{-1}$ within each group is shown in Fig. 1.10. Although the groups separate nicely, their members also show substantial variation in $(\alpha M)^{-1}$. While some of this is estimation error (M is a fairly noisy parameter), some is not. Some of the variation correlates with phylogeny within these vertebrate groups; for example, $(\alpha M)^{-1}$ averages about 0.65, 0.65, 0.30, and above unity for cod, flatfish, brown trout, and some long-lived herring species respectively (Beverton 1963, 1992; Vollestad *et al.* 1993). It behoves us to ask just what sets $(\alpha M)^{-1}$ both within and between the various vertebrate groups. Figures 1.8–1.10 illustrate the power of an approach to life histories which uses dimensionless numbers and looks for invariants. Although the invariance is only approximate, it is strong enough to be of interest.

1.6 Life history theory for the αM number

Why do the groups differ, and why particular $(\alpha M)^{-1}$ values? It will prove more useful to work with αM, the inverse of this number. I believe that the answer to why αM takes on particular values lies in how natural selection acts to set the maturation time itself. In what follows I construct a theory for evolution of the age at maturity and

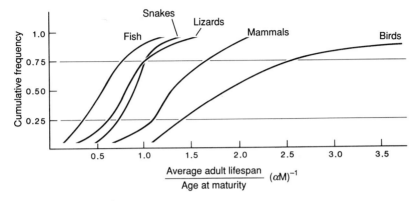

Fig. 1.10 Distribution of the number $(\alpha M)^{-1}$ on a per species basis. Although the various vertebrate groups separate nicely, there is still substantial within-group variation; life history evolution theory aims to account for both within and between-group variation. Data for fish from Beverton and Holt (1959), reptiles from Shine and Charnov (1992), mammals from Millar and Zammuto (1983), and birds from many sources.

require as its output the αM number. In this introductory chapter a phenomenological model is used to illustrate how life history theory can be used for this and how trade-off symmetries play an important role here. Chapters 4 and 5 develop more biologically realistic life history models for αM (and other dimensionless numbers).

I make use of the ESS result (eqn. (1.3)). Now, suppose that $Z(x)$ is the instantaneous mortality rate at age x; in general $Z(x)$ will decrease with x (and may often reach some low and near constant value before maturation although it may go up again late in life (Chapter 7)). We can thus write

$$S(\alpha) = \exp[-\phi(\alpha)] = \exp[-\int_0^\alpha Z(x)\,dx]$$

so that

$$\phi(\alpha) = \int_0^\alpha Z(x)\,dx \quad \text{and} \quad \frac{\partial\phi(\alpha)}{\partial\alpha} = Z(\alpha)\ .$$

The ESS α exists where

$$\frac{\partial\log_e V(\alpha)}{\partial\alpha} = \frac{\partial\phi(\alpha)}{\partial\alpha}$$

which reduces to

$$\frac{\partial \log_e V(\alpha)}{\partial \alpha} \ - \ Z(\alpha) \ .$$

However, if mortality does not change much after maturation, $Z(\alpha)$ is also the adult instantaneous mortality rate M. The ESS equation can thus be written as

$$\frac{\partial \log_e V(\alpha)}{\partial \alpha} \ = \ M \ . \tag{1.4}$$

This equation for the ESS age at maturity is the first step towards obtaining a value for the αM number. The key apparently lies in the $V(\alpha)$ function; eqn. (1.4) does not require that we know the actual $V(\alpha)$ function, only its proportional change with α, i.e. its shape. For example, suppose that we guess that $V(\alpha) \propto \alpha^d$; $V(\alpha)$ is a power function in α with exponent d. Then

$$\log_e V(\alpha) \ = \ \text{constant} \ + \ d \log_e \alpha \ ,$$

and

$$\frac{\partial \log_e V(\alpha)}{\partial \alpha} \ = \ d/\alpha \ .$$

If we put this into eqn. (1.4), something rather interesting happens: the ESS is where $\alpha M = d$. Life histories where $V(\alpha)$ can be treated as a power function in α have the property that αM equals the exponent at the ESS. Here is a theory for the αM number: it suggests that fish have quite high exponents and birds have quite small ones. Better still, we know the values of d (at least approximately) to seek. This is a phenomenological model because nothing really informs us as to what determines the d coefficient, only that, whatever it is, it is similar within fish, birds, etc. It seems clear that to go further we must relate d (or something like it) to general models of growth or other developmental processes. Figure 1.11 shows the symmetry involved in the invariance of αM: $\log_e V$ is a linear function of $\log_e \alpha$, and we can move this function up or down, i.e. change its height, without altering the ESS value for αM. As long as the slope, i.e. the shape, does not change, neither does the ESS αM. To repeat the title of this chapter, *invariants*

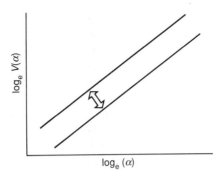

Fig. 1.11 The life history evolution model developed in the text has the trade-off assumption $\log_e V(\alpha) = \text{height} + d \log_e (\alpha)$. The ESS αM number is invariant to changes in height, and is only dependent upon the slope d.

imply deeper symmetries; invariance in αM leads us to search for symmetries or invariance in the trade-offs which control life history evolution. This theme, and its further development, is the core of this book, particularly in Chapters 4 and 5.

1.7 Allometry

One additional approach to life history invariants is sufficiently distinct to deserve separate mention. Within a taxon, such as mammals, a great many physiological, life history, ecological, and size variables are power functions of adult body size W (something $\propto W^P$) as illustrated in Fig. 1.12. These allometric relations enjoy a vast literature including books by Peters (1983), Calder (1984), Reiss (1989) and Harvey and Pagel (1991, Ch. 6). It has been known for decades that the exponent P tends to take on characteristic values for different kinds of functions (Stahl 1962; Linstedt and Calder 1981; Lavigne 1982; Calder 1984; Damuth 1987; Linstedt and Swain 1988). For example, in across-species plots whole-body rates such as metabolism, production, ingestion, ventilation, blood flow, and so forth scale with $P \approx 0.75$. Capacities or size variables such as litter weight, body size at weaning, heart volume, and so forth tend to have $P \approx 1$. Finally, timing variables such as cycle lengths, age at maturity, life-span, and yearly clutch size have $P \approx \pm 0.25$. Within these three classes we can construct a vast array of approximate body size invariants since $W^P W^{-P} \propto W^0$. This has been done by Stahl (1962) and many more recent authors (e.g. Lavigne 1982; Linstedt and Swain 1988).

As a concrete example consider Fig. 1.13 where the fitted relations of $\log_e b$ and $\log_e \alpha$ versus $\log_e W$, where W is the adult weight, are

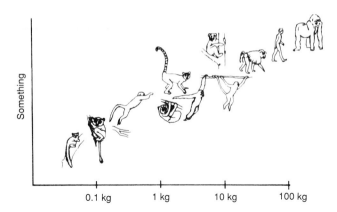

Fig. 1.12 Life history variation is well predicted by adult body weight across primates (and other mammals); 'something' refers to the timing of crucial events, rates, or size relations in the life history, such as yearly clutch size, body size at weaning, gestation, weaning, juvenile period, or life-span. Relationships are typically linear when both the adult weight and timing axes are logarithmically scaled. Three slopes (0.25, 0.75, 1.0) are common in the data; ratios of variables with the same slope will be invariant with body size (i.e. $W^P W^{-P} \propto W^0$). After Harvey and Nee (1991); diagram reproduced with permission from Fleagle (1985).

plotted for the two groups of mammals discussed earlier (Fig. 1.3, primates compared with other mammals). All lines have slopes near $+0.25$ (α) or -0.25 (b). It should be noted that the primates fall on very different lines from the other mammals; for a fixed body size, primates have *much* higher α values and *much* lower b values. Since both groups have $\alpha \propto W^{0.25}$ and $b \propto W^{0.25}$, both will have constant αb. It should also be noted that the respective $\log b$ and $\log \alpha$ lines cross each other at almost the same height on the y axis; this means that both groups share the *same* αb number, as originally displayed in Fig. 1.3. The two groups differ greatly in the heights of the scaling relations, but not in the αb number.

One aim of this book, particularly for mammals (Chapters 5 and 6), is to use life history theory to predict the numerical values of some of these body size invariants (e.g. αM, αb). Since the invariants are the ratio (product) of allometries, this requires that we generate the allometries themselves from more basic processes. The literature abounds with arguments as to why P should take on particular values. More rarely, attempts are made to predict the heights of the allometries even though the above invariants are given by the ratio (product) of the heights. Only by understanding the heights of the allometries can we understand the puzzle presented in the last paragraph and Fig. 1.13:

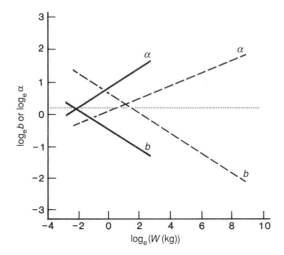

Fig. 1.13 Allometries for yearly clutch size b (in daughters) and age α at first reproduction for 72 primate species (—) and 40 species of other mammals (- - -). Lines have slopes near $+0.25$ (α) or -0.25 (b) and correlations in the range 0.75–0.90. The value of the number αb is the same for both groups (original data plotted in Fig. 1.3) as shown by the fact that the corresponding log b and log α lines cross at the same height on the y axis (dotted line). Data for primates from Ross (1992) and for non-primates from Hennemann (1983).

αb is the same while the heights of its component allometries (b, α) differ greatly for the two groups of mammals.

Allometries can of course be multiplied or divided by each other to make up new allometries; the W^0 discussed above is a special case. This is often a useful sort of manipulation (e.g. Calder 1984). My approach to mammals (Chapter 5) involves assuming a 0.75 allometry (individual production) as a basic input and combining it with life history theory (and population stationarity) to derive formulae for other allometries; I make no attempt to explain the 0.75 production scaling itself. I believe that it represents a fundamental coupling between an individual organism and its environment, but my attempts to derive it from even more basic considerations have thus far failed. Most researchers concerned with 0.75 allometries have focused upon explaining metabolism (e.g. McMahon 1973; critically reviewed by Peters 1983). Metabolism used in this way may well prove a red herring—a convenient measurable which merely parallels the scaling of production (personal growth and offspring production). While many have suggested that metabolic rate ought to be important in influencing life history variation, recent careful work with data from both birds and mammals does not support any such direct connection within these two taxa (Trevelyan *et al.* 1990; Harvey *et al.* 1991). Natural selection ought to be concerned with self and offspring production, not simply heat output. Such is the approach adopted here; production is taken as the basic input to the life history questions.

Finally, I make some comments on statistical and scientific methods for allometry. With reference to Fig. 1.14, it is well appreciated that different taxa sometimes show different heights for allometries of the same variable. For example, birds have longer maximum life-spans than mammals of the same body size (Western 1979; Western and Ssemakula 1982). I show this for two −0.25 scalings (A and B) in Fig. 1.14. However, I have made the scatter (precision) different for the A and B allometries. A gives a much better fit. We can all agree that A and B have different heights, but would we agree that both slopes are −0.25? It depends upon statistics. I have drawn the figures so that a regular regression, which puts all the error in the *y* direction, would yield slopes of −0.25. LaBarbera (1989) would criticize this statistical assumption and insist that the appropriate statistical model would distribute error more evenly on the two axes, for example major axis regression (Ricker 1973) which assumes equal error variance for the *x* and *y* axes. These statistical niceties can make a difference when the allometry is not perfect, as in B in Fig. 1.14. Damuth (1987) claimed a −0.75 scaling for

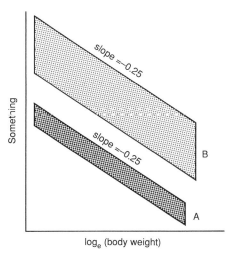

Fig. 1.14 Scatter plots for two −0.25 life history allometries; A has about three times the precision of B. The text discusses why regular regression is probably appropriate here. It also suggests how one might begin to account for the scatter around the log lines.

population size (density) in mammals, LaBarbera (1989) used Damuth's data to claim a −1 scaling; these differences depend upon the statistical model. LaBarbera (1989) would find different slopes for A versus B in Fig. 1.14. Many researchers have realized that the appropriate regression model follows from the way that error variance is distributed on the *x* versus *y* axes (Ricker (1973) gives a good discussion). In this allometry case I believe that Damuth is correct and LaBarbera is incorrect, and in general I believe that regular regression is appropriate for ecological and life history allometry. Three reasons, only one of which is statistical, lead me to this suggestion.

First, body size (the *x* axis) must generally be known with much less error than most life history variables (e.g. mortality rates). Second, variation in the *y* direction is not all measurement error. Harvey and his colleagues (Harvey and Pagel 1991; Chapter 5 of this book) have shown that the residuals from many and various life history allometries are correlated with each other. This would not be expected if deviations from the scaling line were all due to measurement error. Third, treating the imprecision of an allometry as representing mere noise (B versus A in Fig. 1.14) ignores the more basic question of what generates the allometry in the first place. Proposed basic generators of life history allometries will be considered in Chapter 5; one of that chapter's main messages is that imprecision results when the species in the data set share similar but not identical parameter values for the causal factors which generate the allometries themselves. This approach can explain Harvey's correlations between residuals. It also offers a pathway to explain or at least pose hypotheses about why a particular species (family, order) deviates from the scaling line (Wooten 1987). Again, the question is not statistical, about the fitting of allometries, but how to understand allometry as the product of more basic processes. Peters (1983, 1991) argues that this approach is a complete waste of time and is bound to fail. I could not disagree more strongly with the purely descriptive approach advocated by Peters for the study of allometry.

1.8 Phylogenetic methods

Except for mammals (Berrigan *et al.* 1993), *detailed* phylogenetic information in testing life history hypotheses (Harvey and Pagel 1991) is ignored in this book; *in particular, historical transitions are ignored.* My reasons are threefold. First, I assume stabilizing natural selection so that the maintenance of the characters is the independent evolutionary event of interest; species (even populations) are independent data points under the normalizing selection assumption. Two species close in phylogeny may be called similar because of their common history, but for this book they are similar because of similar trade-offs. I realize that phylogenetic information may help us guess at the commonality of the trade-offs (see Fig. 1.9) since closely related species will often share similar ecologies, and thus similar selective forces in the production of their life histories. However, the aim is to understand the common trade-offs. My second reason is that, while acknowledging the usefulness of some historical information (if the theory says y leads to x, but historically x comes first, the theory is destroyed), I believe that such an approach focuses attention on the least interesting part of the problem (historical reconstruction) and away from the most interesting part (to me anyway)—the attempt to build theories to predict evolutionary equilibria, and to understand where trade-offs come from and how they guide evolution. The reconstruction of history becomes much more interesting when the estimated phylogenies are used to search for general rules about history itself, independent of the particular phylogenies (Nee *et al*. 1992). Third, life histories are usually defined by continuous rather than discrete characters; historical predictions (the direction of evolution) are very difficult for multidimensional continuous characters, while equilibrium or ESS predictions are much more straightforward, if not exactly simple.

1.9 Book layout: a short summary

In Chapter 2 we examine the sex ratio from the viewpoint of the one mother–one father symmetry, often combined with other asymmetries that result in non-equal ESS sex ratios. The idea that asymmetric (symmetric) ESS sex ratios are a direct result of asymmetric (symmetric) shapes in certain trade-off relations is discussed. Use is also made of the stationary population assumption—the SAD assumption (Fig. 1.4). In Chapter 3 we look at alternative male life histories (under the SAD assumption) and cast them in the same framework as the sex ratio, deriving ESS results that combine symmetric and asymmetric parts. In Chapter 4 we examine organisms where growth continues after maturation. We consider dimensionless numbers made up from timing and body size variables, and develop life history theory to explain their occasional invariance. Chapter 5 is a detailed development of life history allometry for female mammals, again using timing and size variables. Life history theory is asked to output the invariants; in this chapter we also look at what causes the invariance in αM to be slightly broken within mammals. We also look at body size asymmetry for males and females. Chapter 6 is a departure from the rest of the book in that it does not develop life history theory but uses the output of that theory (mostly from Chapter 5) to show what body size and population size invariants can tell us about some population dynamic processes, particularly body size allometry for a population's maximum intrinsic rate of increase. Chapter 7 is about ageing or senescence, an evolutionary result of the fitness asymmetry inherent in age itself (where older is worth less, in a fitness sense, than younger). Finally, Chapter 8 points to the future through a discussion of unsolved and partially solved problems (and possibilities) that have emerged throughout the book.

Finally, as also noted in the preface, the chapters are not sequential. I suggest reading the following three groupings in any order: Chapters 2 and 3; Chapter 4; Chapters 5 and 6. Read Chapters 7 and 8 last. Be prepared to be sent to another chapter from time to time.

2

Sex allocation

2.1 Introduction and overview

Darwin (1871) first raised the question of sex ratio evolution, and saw it as both important and enigmatic. However, he was unable to make much headway with the problem and declared it a puzzle for the future. This particular future arrived about 60 years later, when Fisher (1930) pointed out for diploid dioecy that under autosomal inheritance half the genes passed to zygotes come from males and half come from females. The transformation involved in this inheritance symmetry can be thought of as an exchange of labels in following ratio:

$$\frac{\text{number of genes to zygotes from sex } i}{\text{number of genes to zygotes from sex } j}.$$

The value of this ratio is unity, independent of which sex is assigned to sex i. The ratio is invariant to the exchange of labels. (A comparable but more complex invariant applies to out-crossed haplo-diploid inheritance). Fisher noted that this one mother–one father symmetry generates frequency-dependent natural selection on sex ratio, resulting in an evolutionary equilibrium (ESS) where half the reproductive resources are devoted to daughters and half to sons.

Although interest in sex ratio as a phenotypic trait did not become widespread for another 35 years, it is difficult for us to overestimate the importance of Fisher's brief and characteristically cryptic remarks. Almost all the innovations in thinking about sex ratio can be viewed as alterations of assumptions implicit in Fisher's scheme (Bull and Charnov 1988). Sex ratio theory is a well-defined microevolutionary theory that makes striking testable predictions, some of them non-intuitive. As with most evolutionary life history theory, we generally focus on the equilibrium (or ESS) predictions. However, for evolutionists who distrust such static results, opportunities abound for looking at the short-term dynamics of sex ratio phenotypes under a detailed knowledge of inheritance patterns.

Of course, there have been substantial conceptual additions to Fisher's ideas as well as an immense amount of empirical work (for review see Nonacs (1986), King (1987), and Wrensch and Ebbert (1993) for insects, Clutton-Brock and Albon (1982), Clutton-Brock and Iason (1986), Hrdy (1987), and Frank (1990) for mammals, and Clutton-Brock (1986), Frank (1990), and Gowaty (1991) for birds). Empirical reviews of hermaphrodites and sex changers, concepts to be developed below, include those by Freeman *et al.* (1980), Charlesworth and Charlesworth (1981), Policansky (1982), Lloyd (1984), Lloyd and Bawa (1984), Goldman and Willson (1986), Charlesworth and Morgan (1991), and Brunet (1992) for higher plants, and by Warner (1988*a*,*b*) for fish. The field of sex allocation is surveyed in two symposia volumes (Policansky 1987; Mangel 1990) and by Charnov (1982).

As an overview and introduction it is useful to divide advances since Fisher into six categories. After a brief discussion of each and a terse overview of some general ESS allocation results, the broad theme is developed that biased population-wide sex ratios (or allocations) can *often* be understood as representing Fisher's inheritance symmetry combined with asymmetric opportunities to reproduce through male versus female function. The degree of bias, or deviation from equality, is usually a direct reflection of the degree of asymmetry. We shall apply this idea to hermaphroditic plants, sex-changers, and nematodes with environmental sex determination.

One substantial addition to Fisher's ideas is the realization that one mother–one father symmetry generates the same kind of frequency dependence of fitness for allocation to male and female function in hermaphrodites (Charnov *et al.* 1976; Charnov 1979*a*,*b*; Charlesworth and Charlesworth 1981; Lloyd 1984), and the allocation of the reproductive life span between male and female function in sex-changers (Ghiselin 1969; Warner 1975*a*,*b*; Warner *et al.* 1975; Leigh *et al.* 1976; Charnov 1979*a*,*c*). These allocation problems, together with the stability of the various gender types themselves (e.g. dioecy versus hermaphroditism), define the field of sex allocation (Charnov 1982); all the factors discussed below potentially apply to these gender types, in addition to the sex ratio under dioecy.

The second addition to Fisher's ideas is the development of very general ESS techniques to solve sex ratio problems (Maynard Smith 1982). Shaw and Mohler (1953) pioneered the ESS approach by looking at the relative fitness of a rare mutant altering sex ratio, while MacArthur

(1965) showed that at the ESS natural selection favours females that control their clutch size, sex ratio, and allocation of resources among offspring so as to maximize the product of the reproductive gains through sons multiplied by the gains through daughters. Hamilton (1967) extended the ESS technique to spatially structured populations. Shaw–Mohler type equations and product theorems for the ESS are now ubiquitous in sex allocation work (e.g. Stubblefield 1980; Frank 1985, 1986*a,b*, 1987; Charnov and Bull 1986; Charlesworth and Charlesworth 1987; Lloyd 1987). Karlin and Lessard (1986) have introduced very powerful mathematical techniques for the study of sex ratio evolution. However, their deeply genetic approach has yet to lead to any major empirical advance, and ESS techniques remain the approach of choice for most workers.

The third major addition to Fisher's ideas is the recognition that violation of his one mother–one father symmetry can lead to biased sex ratios. Sex linkage, and other forms of non-Mendelian inheritance, may give rise to such violation (Shaw 1958; Hamilton 1967; Werren *et al.* 1988); haplo-diploidy with inbreeding certainly does (Hamilton 1972, 1979; Frank 1983, 1985, 1986*a,b*). Nuclear genes are not the only heritable elements potentially involved in sex ratio determination. A variety of extrachromosomal elements (ECEs) or heritable infections with parent-to-offspring transmission are known from insects and many plants (Gouyon and Couvet 1987; Werren *et al.* 1988; Ebbert 1993). Since the elements rarely show symmetric transmission by sex of offspring (or gender), the element's favoured sex ratio is almost always biased (mostly towards daughters) and is usually in conflict with the sex ratio favoured by the autosomal nuclear genes. Indeed, some populations of the parasitoid wasp *Nasonia* have both son and daughter-transmitted ECEs, so that the elements are sometimes even in conflict with each other over sex ratio (Werren *et al.* 1981, 1986, 1987; Skinner 1982, 1985, 1987; Huger *et al.* 1985; Werren and van den Assem 1986; Werren 1987*b*). These ECEs allow us to ask a great variety of evolutionary questions, an example being the extent of co-evolution between the respective 'genetic parties' (Ebbert 1993).

The fourth addition refers to violation of a symmetry implicit in Fisher's ideas—the assumption of proportional gains through sons versus daughters. This hidden symmetry assumption was (is) that as the mother shifts resources from one sex to the other sex, the ratio of her reproductive gains to losses is a constant and is the same value for all mothers (or circumstances). Three types of violations of this symmetry have been well characterized: asymmetric gains caused by (1) population structure (e.g. sib mating), (2) individual circumstance, or (3) time of year.

Local mate competition (LMC) commonly refers to female-biased sex ratios favoured under sib mating, an extreme form of population structure

(Hamilton 1967, 1979; Werren 1980; Frank 1983, 1985, 1986a,b). The mother is selected to overproduce daughters, since sons in excess of those necessary to inseminate their sisters add nothing to the mother's fitness (an asymmetry). This situation can be generalized to allow partial sib mating, competition among female offspring for resources if not mates (Clark 1978), variable fecundity among the mothers (Werren 1980, 1983, 1984a,b, 1987a) and so forth. These alterations affect the extent and direction of the son–daughter asymmetry and thus the ESS sex ratio bias. LMC is discussed in nearly all the chapters of a recent book on arthropod sex ratios (Wrensch and Ebbert 1993), and insects (and mites) provide a wealth of examples and opportunities to test and refine the LMC ideas. For example, female-biased sex ratios are widely associated with haplo-diploidy in many insect and mite groups. Haplo-diploidy and inbreeding favour biased sex ratios independent of LMC because inbred mothers are more related to daughters than to sons (a second violation of the Fisher symmetry); inbreeding does not have this affect under diploidy. Therefore, inbreeding and LMC interact to favour female-biased sex ratios under haplo-diploidy but not under diploidy. LMC-type models also apply to inbreeding systems in hermaphrodites (Charnov 1979b; Charlesworth and Charlesworth 1981) and are particularly well studied in self-fertilizing and partially self-fertilizing plants (Charnov 1987; Cumaraswamy and Bawa 1989; Charlesworth and Morgan 1991). Slightly altered model predictions have been developed for out-crossed hermaphrodites who mate in small groups of perhaps two or three individuals (Charnov 1980: Fischer 1981, 1984). In this case there are supporting data from various fish (Petersen 1990) and barnacles (Raimondi and Martin 1991).

The second example of asymmetric gains (individual circumstance) refers to sex ratio alteration by individual mothers as a function of the quality of the son or daughter who would be produced in a particular context (Trivers and Willard 1973 (mammals); Charnov and Bull 1977; Charnov 1979a; Bull 1981; Frank 1987). The most obvious example is the host size effect in parasitoid wasps (Charnov *et al.* 1981). Small hosts produce small adults, large hosts produce large adults, and size is thought to affect male (son) versus female (daughter) fitness differentially (asymmetrically). Some laboratory data support the idea that females gain more than males from larger size (van den Assem *et al.* 1989). Correspondingly, daughters are often overproduced in large hosts and sons are overproduced in small hosts. This sex differential fitness effect may also be widespread among hermaphrodites (Charnov and Bull 1977); it is also considered a key selective force in sex change (Warner 1975a,b) and in the evolution of environmental sex determination where an individual chooses its gender depending upon its developmental opportunities

(Charnov and Bull 1977, 1989*a*,*b*; Bull 1983; Bull and Charnov 1989). In these instances it is the individual's ability to reproduce through male versus female function related to its state (small/large, young/old, and so forth) which favours the conditional sex expression.

Thirdly, seasonal variation in the opportunities for son versus daughter reproduction may also select for adjustment of sex ratio, here with respect to the time of year. The theory and data are discussed by Werren and Charnov (1978) and Seger (1983); a long-term study has recently been performed by Brockmann and Grafen (1992).

LMC and the Trivers–Willard (1973) model for individual sex ratio adjustment in mammals are the most studied of all sex ratio theories, perhaps the most studied of all ESS theories. While I have discussed violations of the two Fisher symmetries (inheritance and proportional gains through each sex) in a piecemeal fashion, it is clear that more than one kind of violation may be present in any particular situation (e.g. Werren 1984*a*,*b*; Werren and Simbolotti 1989; Brockmann and Grafen 1992).

The final two major additions to Fisherian thought are the interaction of sex ratio and kin selection in the social insects (Hamilton 1964; Trivers and Hare 1976), and the interaction of sex ratio with the sex-determining mechanism.

Kin selection and sex ratio come together in the social Hymenoptera. Trivers and Hare (1976) pioneered work in this field by noting that female workers (which are three times as related to their sisters compared with their brothers, a large reproductive asymmetry from the worker's perspective) ought to prefer a resource investment of 3:1 if they rear sisters and brothers (respectively) as reproductives. The queen prefers the usual Fisherian ratio of 1:1. Testing and elaboration of these worker–queen investment conflicts has become a large area of study within work on social insects (Nonacs 1986; Seger 1991*a*,*b*).

Sex ratio selection plays a major role in selection for, or against, various forms of sex determination, both on a microscale such as alternative forms of heterogamety (Bull and Charnov 1977) and on the macroscale of questions like environmental versus genotypic sex determination (Bull 1983). Conversely, the sex-determining mechanism may enhance (e.g. haplo-diploidy) or greatly constrain (e.g. sex chromosomes) short-term alteration of sex ratio. Facultative alterations of sex ratio in birds and mammals are usually characterized by small shifts away from 1:1, and the literature contains many discussions of how birds and mammals can manage this while having sex chromosomes (cf. reviews cited earlier).

Of the six major categories of addition to Fisherian thought, one is mostly technical (use of explicit ESS techniques), two refer to violations of his one mother–one father symmetry (e.g. extrachromosomal elements

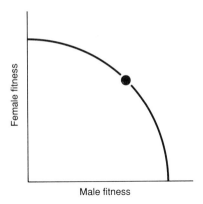

Fig. 2.1 Sex allocation. The figure shows a hypothetical trade-off between male fitness and female fitness. The axes can be labelled in any of a number of ways. For example, they may be son production versus daughter production, the problem of interest being selection on the sex ratio. Other interpretations (see text) include reproductive gains for time as a male or female in a sex-changer, or reproductive gains for pollen versus seed in a hermaphroditic plant. For all of these seemingly different cases, the equilibrium favoured by selection takes the same form, shown by the full point on the trade-off curve. This point marks the value which maximizes the product of the gain through male fitness multiplied by the gain through female fitness.

affecting sex ratio; asymmetric relatedness in the social insects), one refers to the use of that symmetry to understand sex allocation in hermaphrodites and sex-changers, and one refers to the inheritance symmetry combined with asymmetric opportunities for reproduction through male (son) versus female (daughter) function (e.g. local mate competition).

In the rest of this chapter we study sex allocation under the assumption that everyone has one mother and one father.

2.2 Sex allocation under the Fisher inheritance symmetry

The problem of sex allocation can often be visualized as in Fig. 2.1. The axes are fitness through male function and through female function. The curve represents the possible trade-offs, with the end-points representing all-male function and all-female function. We consider three cases.

(i) In a species that is dioecious, the curve represents sons and daughters, and we are concerned about the evolution of the sex ratio.

(ii) In a species that is a sex changer, the axes represent reproductive gain as a male versus reproductive gain as a female (the trade-off curve then represents various proportions of the life-time spent as a male versus a female).

(iii) In a species that is a simultaneous hermaphrodite (consider it a plant), the axes represent reproductive gain via seeds versus pollen.

Using population genetic arguments for autosomal genes, it can be shown that the equilibrium (ESS) favoured by natural selection is often that value on the trade-off curve which maximizes the product of the fitness through male function multiplied by the fitness through female function, a generalization of MacArthur's (1965) product theorem for the sex ratio (Charnov 1979a). The trade-off curve is assumed to be linear or convex (bowed out); some cases where it is not are developed below.

The discussion here will focus on the Fisher inheritance symmetry (autosomal genes) combined (often) with asymmetric gains through the investment of resource in male versus female function. The hermaphroditism section is wholly theoretical, developing the flavour of the symmetry approach. In the cases of dioecy and sex change we examine data where some male–female gain asymmetries result in biased population-wide sex ratios.

2.3 Simultaneous hermaphroditism

Consider a perfect flowered plant, where each flower contains male and female reproductive elements. For an out-crossed species consider the reproductive consequences, through both the male m and female f functions, for the shifting of resources from all male through various degrees of hermaphroditism to all female. The curve shown in Fig. 2.2 is the boundary of all possible types. It is easy to show (Charnov *et al.* 1976) that such a convex curve implies that out-crossed hermaphroditism is stable; a concave curve favours dioecy. The boundary between a convex and a concave trade-off is the straight line $m + f = 1$. The intuitive reason that 'convexity favours hermaphroditism' is based on the view that a pure female (male) giving resource to a male (female) function gains more reproductive success than it loses. In the most extreme case possible, the hermaphrodite would be almost as good a male as the pure male and would produce almost as many seeds as a female. To visualize this, imagine pushing the convex curve of Fig. 2.2 outward (end-points still at $m = 1$, $f = 1$) until the dot is at the co-ordinates (1,1). What happens if the trade-off curve is concave–convex is illustrated in Fig. 2.3; this will often lead to the ESS being a mixture of hermaphrodites and a pure sex.

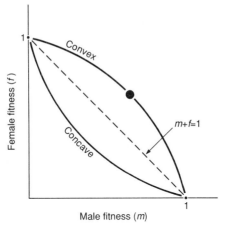

Fig. 2.2 Two possible trade-offs between male (pollen = m) and female (ovules, seeds = f) fitness for a hermaphrodite. Pure males and females (end-points) are assigned a fitness of unity; the hermaphrodite's fitness through each gender function is scaled relative to the pure sex. For out-crossed hermaphroditism to be stable, the trade-off must be convex. For this, the equilibrium allocation of resources to male versus female reproduction is the point which maximizes the mf product (full point).

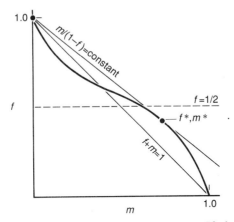

Fig. 2.3 An ESS which is a mixture of sexual types. If the curve is convex–concave, it may be possible for a pure sex (in this case a female) to invade the hermaphrodite population. The resulting mixture, here hermaphrodites and pure females (called gynodioecy), is evolutionarily stable if three conditions are met. First the hermaphrodite's allocation to male versus female function is at point m^*, f^*; this is the place on the trade-off curve which maximizes $m/(1-f)$. Imagine drawing a line out from the point $f = 1$, $m = 0$ and rotating it upwards until it intersects the most outward possible point on the trade-off surface; this yields f^*, m^* as shown in the figure. The second and third conditions for gynodioecy to be an ESS are $m^* + f^* > 1$ and $f^* < 0.5$. From Charnov *et al.* 1976.

The rest of the hermaphrodite discussion will focus on the roles of a particular shape symmetry. Suppose that a proportion r of reproductive resources is given to the male function ($1 - r$ to female). Let male reproductive success, relative to a pure male, be given by $m = r^{n_1}$; female reproduction is given by $f = (1 - r)^{n_2}$. The exponents (n_1, n_2) control the *shapes of the gain curves* (Charnov 1979b); Fig. 2.4 illustrates this for $f = 1 - r$ and various n_1 values. Provided that hermaphroditism is stable, the ESS r maximizes mf (or maximizes $r^{n_1} (1 - r)^{n_2}$). Setting $\partial \log_e (mf)/\partial r = 0$ yields $r/(1 - r) = n_1/n_2$. The ESS sex allocation is an exact reflection of the shapes of the two gain curves. If they are symmetric ($n_1 = n_2$), the male-to-female allocation will be 1:1; otherwise it will be biased towards the function with the largest exponent. Asymmetry ($n_1 \neq n_2$) in the shapes of the gain functions is reflected in asymmetric allocation to the sex functions. Figure 2.4 illustrates this for $n_2 = 1$ and various n_1 values.

However, will hermaphroditism itself be stable? Other possibilities include dioecy and populations consisting of mixtures of a hermaphrodite morph and a pure sex (androdioecy (male) or gynodioecy (female) (Fig. 2.3)). As argued in Fig. 2.3, mixtures can be stable if the male–female trade-off relation (m versus f) is concave–convex. Figure 2.5(a) shows the ESS outcomes for various n_1 and n_2: both less than unity

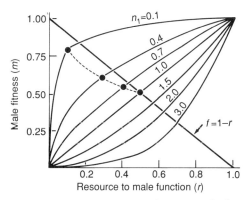

Fig. 2.4 If r is the proportion of resources given to male function, male fitness m is given by $m = r^{n_1}$. The graph shows male fitness (the male gain curve) for various values of n_1. Female fitness is assumed to be proportional to the resource input into eggs, i.e. $f = 1 - r$. Hermaphroditism is stable to dioecy only if $n_1 < 1$, i.e. only if male reproductive success m saturates or shows a law of diminishing returns with resource input. The full points show the equilibrium (ESS) allocation r to male function for 4 n_1 values. The ESS result is $r = n_1/(1 + n_1)$, so that as n_1 decreases, so does the ESS r. Further discussion in the text. From Charnov 1979b.

implies hermaphroditism; both greater than unity implies dioecy. If one is less than unity and the other is greater than unity we obtain a variety of outcomes depending upon the magnitude of the asymmetry. The proportion of the pure sex in the mixed populations is indicated in Fig. 2.5(a). Figure 2.5(b) shows the hermaphrodite's ESS allocation *r* to the male function; it should be noted that with androdioecy (gynodioecy) the hermaphrodite morph gives most of its resource to the female (male) function.

Figure 2.5(a) shows an interesting puzzle if we consider the following thought experiment. Set the female exponent, n_2 equal to 0.5 and gradually increase the male exponent n_1; then the male function gives greater returns per unit investment at higher n_1. As n_1 goes above 1.5, hermaphroditism becomes unstable but the pure sex that invades is not the male but the female. This may seem paradoxical since it was increasing returns on investment in male function that were going up. The solution to the puzzle can be found in Fig. 2.5(b); as n_1 increases, the ESS sex allocation within the hermaphrodite is more towards male function. A female morph and not a male invades at $n_1 = 1.5$ because the female has large fitness gains through the reallocation of the greater amount of resource that was going to the male function. Therefore, in a mixed population, the pure sex will have $n < 1$ (diminishing returns on investment), while the hermaphrodite morph will have increasing returns on investment in the other sex function.

As a final example for hermaphroditism, consider a plant that invests its pool of resources into pollen, seeds, and structures for attracting pollinators. Charnov and Bull (1986), Charlesworth and Charlesworth (1987), and Lloyd (1987) have developed some cases here. They include factors such as seasonality of resource availability, partial self-pollination, attraction of seed dispersal (fruit-dispersal) agents as well as pollen vectors, and so forth. Attention is restricted to the simplest out-crossed situation (also please consult the original papers for the side conditions necessary for hermaphroditism itself to be stable). For an individual let *m* be the number of pollen grains, *f* the number of seeds, and Q_m and Q_f the success per pollen grain and seed respectively, due to attraction of pollen vectors. The ESS allocation for the hermaphroditism now maximizes as follows (Charnov and Bull 1986): $(mQ_m)(fQ_f)$. Let *x*, *y*, and *z* be the proportions of the limiting resource given to attractant, seed, and pollen respectively. Assume that

$$m \propto z^{n_1} \qquad Q_m \propto x^{n_3} \qquad f \propto y^{n_2} \qquad Q_f \propto x^{n_4} \; .$$

The product now becomes $z^{n_1} y^{n_2} x^{(n_3 + n_4)}$. By the method of Lagrange multipliers, with the constraint $z + y + x = 1$, the ESS allocation for

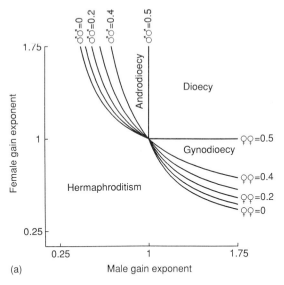

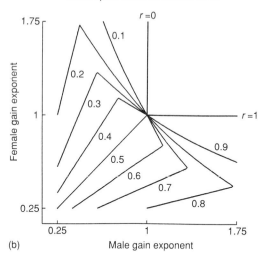

Fig. 2.5 (a) ESS sex types for the model $m = r^{n_1}$ and $f = (1 - r)^{n_2}$ for various values of n_1 (male exponent) and n_2 (female exponent). Gynodioecy (andro-dioecy) is a mixed population of females (males) and hermaphrodites. If both exponents are less than unity, only hermaphroditism is stable; if both are greater than unity, only dioecy is stable. If one is less than unity and the other is greater than unity, we obtain a variety of outcomes depending upon the degree of *asymmetry* in the exponents. The proportion of the mixed populations made up of the pure sex is also indicated. (b) The ESS sex allocation r within the hermaphrodite

the relative allocations $x:y:z$ is found to be in the ratios $n_3 + n_4$: $n_2:n_1$. As before, the *shapes* of the gain functions control the allocation ratios. This often remains true even for more complex cases (e.g. partial self-pollination), as developed in the papers cited above.

2.4 Sex reversal: breeding sex ratio

Sex reversal comes in two forms, protandry (male first) and protogyny (female first). Protandry is widely scattered among invertebrates (and a few fish) while protogyny is very common among coral reef fish. Ghiselin (1969) suggested that natural selection ought to favour sex change with increasing size (or age) if one sex reproduces better while small whereas the other reproduces better when large. Warner (1975a), Warner *et al.* (1975), and Leigh *et al.* (1976) used an ESS approach and altered Ghiselin's suggestion as follows: sex reversal is favoured where one sex gains relatively more reproductive ability with age or size. While age-specific fertility is usually considered to be the ability which changes differentially (asymmetrically), there are also versions of the theory which examine sex-differential advantage in mortality and/or growth (Charnov 1982; Policansky 1982; Iwasa 1991).

This discussion will be limited to the fertility advantage version and the aim is to discover what determines the adult or breeding sex ratio. Consider a large population with overlapping generations which is numerically stable (through density-dependent immature mortality) and has a stable age distribution. Define the following for protandry. $b(x)$ is the birth rate of an age x female relative to the birth rate of an age z female, where z is an arbitrary age chosen simply to make female fertility a relative measure. $Q(x)$ is the fertility of an age x male relative to the fertility of an age z male. This age z is the same as used above, so that an individual operating as a male or a female at age z has a relative fertility of unity. $Q(x)$ is the relative ability of an age x male to fertilize eggs (e.g. compete for females). $L(x)$ is the probability that an

---

Fig. 2.5 (*continued*)
morph. Androdioecious populations have $r < 1/2$ while gynodioecious populations have $r > 1/2$; r always favours the opposite sex from the pure sex. Of course, with only hermaphrodites present, the ESS is given by

$$\frac{r}{1-r} = \frac{\text{male gain exponent}}{\text{female gain exponent}}.$$

Reproduced with permission from Seger and Eckhart (unpublished).

individual is alive at age x. This is the usual life table definition from demography.

For this stationary population we can designate the genetic contribution of an individual (changing sex at age τ) through male function as

$$m(\tau) = \int_0^\tau L(x)Q(x)\,dx \quad . \tag{2.1}$$

Female lifetime fitness is given as

$$f(\tau) = \int_\tau^\infty L(x)b(x)\,dx \quad . \tag{2.2}$$

The ESS age of sex change $\hat{\tau}$ maximizes $m(\tau)f(\tau)$; setting $\partial \log_e(mf)/\partial\tau = 0$ yields

$$\frac{b(\hat{\tau})}{\displaystyle\int_{\hat{\tau}}^\infty L(x)b(x)\,dx} = \frac{Q(\hat{\tau})}{\displaystyle\int_0^{\hat{\tau}} L(x)Q(x)\,dx} \quad . \tag{2.3}$$

This equation has the following intuitive meaning. The left-hand side is the fitness of an age $\hat{\tau}$ individual reproducing as a female ($b(\hat{\tau})$ divided by the sum of the b values for all the other females); the right-hand side is the same for an age $\hat{\tau}$ individual reproducing as a male. Age $\hat{\tau}$ is where an individual is indifferent (in fitness gain) to whether it is a male or a female, hence their equalization at the ESS.

The protandry is itself stable if the $m(\tau)$ versus $f(\tau)$ trade-off surface (Fig. 2.1) is convex (bowed out). This is true if $\partial^2 f(\tau)/\partial m(\tau)^2 < 0$; this is satisfied if $b(x)/Q(x)$ increases with x, i.e. if females gain more reproductive ability ($b(x)$) with age than do males ($Q(x)$). This condition is an *asymmetry* with respect to sex (gender) in the way that age (size) maps to reproductive ability. Figure 2.6 shows this condition, with the convention that the heights of the $b(x)$, $Q(x)$ curves are adjusted so that $b(\hat{\tau}) = Q(\hat{\tau}) = 1$; this is allowed since a b or Q curve represents only relative fertility within each sex. We can now rewrite eqn. (2.3) as

$$\frac{\displaystyle\int_{\hat{\tau}}^\infty L(x)b(x)\,dx}{\displaystyle\int_0^{\hat{\tau}} L(x)Q(x)\,dx} = 1 \quad .$$

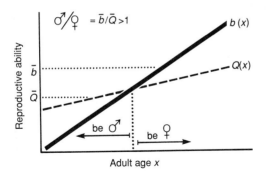

Fig. 2.6 Breeding sex ratio ($\sigma / \female$) under sex change. If both sexes gain a fertility advantage ($\sigma = Q$, $\female = b$) with age, but females gain more (b/Q increases with x), protandry ($\sigma \rightarrow \female$) is favoured and the breeding sex ratio will be σ (first sex) biased. The convention that the relative fertility curves cross at the age of sex change is adopted. The overbar ($\overline{b}$, $\overline{Q}$) means average. See text for detailed development.

If we multiply both sides of this equation by the ratio $\left. \int_0^{\hat{\tau}} L(x)\,\mathrm{d}x \right|$

$\int_{\hat{\tau}}^{\infty} L(x)\,\mathrm{d}x$, we obtain

$$\frac{\displaystyle\int_{\hat{\tau}}^{\infty} L(x)b(x)\,\mathrm{d}x \left/ \int_{\hat{\tau}}^{\infty} L(x)\,\mathrm{d}x \right.}{\displaystyle\int_0^{\hat{\tau}} L(x)Q(x)\,\mathrm{d}x \left/ \int_0^{\hat{\tau}} L(x)\,\mathrm{d}x \right.} = \frac{\displaystyle\int_0^{\hat{\tau}} L(x)\,\mathrm{d}x}{\displaystyle\int_{\hat{\tau}}^{\infty} L(x)\,\mathrm{d}x} . \tag{2.4}$$

It should be noted that the left-hand side of eqn. (2.4) is the ratio of the average b to the average Q ($\overline{b}$ and $\overline{Q}$ respectively), while the right-hand side is, for a stationary population, the ratio of the number of adults younger than age $\hat{\tau}$ to the number older than $\hat{\tau}$, i.e. the adult sex ratio. In a very compact form

$$\sigma / \female = \overline{b} / \overline{Q} . \tag{2.5}$$

Provided that both sexes gain a fertility advantage with age, $\overline{b} > \overline{Q}$ and males should outnumber females at the ESS; the extent of sex ratio bias reflects the asymmetry in the average relative fertilities ($\overline{b}$, $\overline{Q}$) for the sexes, as illustrated in Fig. 2.6.

Under protogyny (female first, male second) we simply reverse the gender labels, so that the general rule is

$$\frac{\text{first sex breeders}}{\text{second sex breeders}} = \frac{\text{average relative fertility of second sex}}{\text{average relative fertility of first sex}} > 1 \; .$$

$$(2.6)$$

Charnov and Bull (1989*a*) discuss how the inequality (first sex more abundant than the second sex) also holds with other selective advantages (growth, mortality asymmetries) leading to sex change. Charnov (1989*b*) generalizes the inequality to androdioecy (males plus protogynous sex-changer); the data are discussed in Chapter 1, Figs. 1.6 and 1.7.

Indeed, the first sex is almost always more abundant under sex change. Figure 2.7 shows data for nine species of labroid fishes (parrotfish and wrasses) from the Great Barrier Reef of Australia and various Caribbean locations. Data for several other protogynous fish are compiled by Charnov and Bull (1989*a*) and Shapiro and Lubbock (1980); in all cases females are more abundant (see also Fig. 1.1). Protandrous species are much less well known (references cited in Charnov and Bull 1989*a*). My own estimates for pandalid shrimp have the first sex (male) making up approximately 55–75 per cent of the breeders; data for the limpet *Patella vulgata* (UK) suggest that the first sex makes up 75 per cent of the breeders. Sex ratio for limpets in the genus *Crepidula* (North America) are often, but not always, biased toward the first sex (Hoagland 1978).

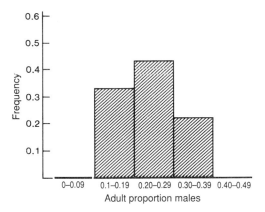

Fig. 2.7 The breeding sex ratio for nine species of labroid fishes. All have an excess of females, the first sex, as predicted by theory. Data from Robertson and Choat (1973), Choat and Robertson (1975), Robertson and Warner (1978), and Warner and Robertson (1978).

2.5 Dioecy: population sex ratio with environmental sex determination

Non-equal primary or birth sex ratios are often associated with LMC and/or violation of the one mother–one father symmetry, the Fisher inheritance symmetry (Bull and Charnov 1988). Here I use symmetry arguments to develop a theory for biased population sex ratios under environmental sex determination (ESD). Through ESD, offspring become male or female in response to one or more environmental factors during development. This is known to occur in fish (Conover and Heins 1987), reptiles (Bull 1980, 1983, Janzen and Paukstis 1991), nematodes (Petersen 1972, 1977), and shrimp (Adams *et al.* 1987, which gives a general review of ESD). Here I concentrate on nematodes.

Although ESD was first reported for nematodes by Christie (1929) in studies of a mermithid parasite of grasshoppers, the most extensive work on this group was done by Petersen and colleagues (Petersen *et al.* 1968; Petersen 1972, 1977) with mosquito parasites. Infective-stage juveniles of these mermithids actively search for and penetrate an individual host larva. Development to adult size proceeds in the host, which is killed when the worms emerge; after one more moult, they become free-living adults. Since adult mermithids do not feed, all reproductive resources are acquired during parasitic development. Some host environments allow more growth, others less; adult worm size (and presumably the amount of reproductive resources) is closely related to the developmental environment. Several ingenious experiments by Petersen have demonstrated conclusively that these parasites exhibit ESD; indeed, ESD seems to characterize the entire Mermithidae family, having been found in every species so far examined (Poinar 1979). Petersen showed that sex expression is closely related to the attainable adult size, with small worms becoming male. Host size, host nutrition level, and parasitic load all affect worm growth opportunities and thus sex expression.

To model this using sex allocation theory define $g(x)$ as the proportion of the emerging adults of body size x, $W_1(x)$ as the relative life-time male fertility of an individual of size x, and $W_2(x)$ as the relative female fertility of a size x individual. Both $W_1(x)$ and $W_2(x)$ are scaled within a sex, so that some arbitrarily chosen size will have both equal to unity; this is analogous to how we defined b and Q for the sex-changer. Provided that W_2/W_1 increases with increasing x (i.e. females gain more reproductive ability with size than do males, a gain asymmetry), all small worms should be male and all large ones female. The threshold τ is the value that maximizes the following product relation (illustrated in Fig. 2.8):

$$\text{maximize} \left[\int_0^\tau g(x)W_1(x)\,\mathrm{d}x \right]\left[\int_\tau^\infty g(x)W_2(x)\,\mathrm{d}x \right] . \qquad (2.7)$$

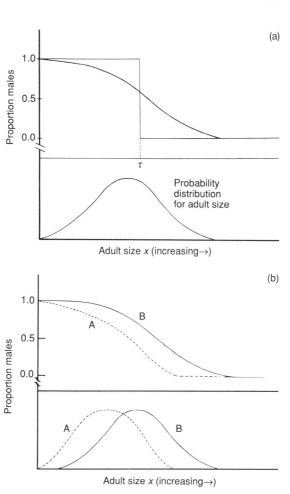

Fig. 2.8 Sex ratio as a function of adult size. (a) provided that females gain more fitness by being larger than males (i.e. $W_2(x)/W_1(x)$ increases with x), the sex ratio is predicted to be female biased among large individuals and male biased among smaller ones. At a threshold size τ, a changeover occurs; however, actual field data would show a more gradual sex ratio shift, and here τ is defined to be the size at a sex ratio of 1:1. (b) If the adult size distribution differs among breeding populations, τ is predicted to differ also. Locations with mostly large individuals (curve B) should show a larger τ when compared with locations with smaller individuals (curve A). Indeed, the entire curve showing the sex ratio as a function of adult size should shift with the adult size distribution.

We can use eqn. (2.7) to make two sorts of predictions. First, the overall emergence sex ratio ought to be male biased. If both $W_1(x)$ and $W_2(x)$ increase with x, and if we simply adopt the convention that $W_1(\tau) = W_2(\tau) = 1$, then the same argument that leads to eqn. (2.5) for the sex-changer gives us the population-wide ESS emergence sex ratio: eqn. (2.7) is analogous to the product theorem for sex-changers. We obtain

$$\sigma/\varphi \; = \; \overline{W}_2/\overline{W}_1 \tag{2.8}$$

where the bar refers to the simple average taken over the body size distribution for each sex. Provided that both males and females gain reproductive ability with increasing size (W_1 and W_2 both increase with x), the population-wide emergence sex ratio will favour males, since $\overline{W}_2$ will always be greater than $\overline{W}_1$. The second prediction is that if we shift the overall size distribution, as illustrated in Fig. 2.8(a), we shall change both $\overline{W}_1$ and $\overline{W}_2$ in the same direction. This means that the overall emergence sex ratio (eqn. (2.8)) will tend to be approximately a constant across the environments brought about through shifts in the ESS τ values. This is really the same sex ratio invariance as shown for a sex-changing fish in Fig. 1.1.

Blackmore and Charnov (1989) have tested these ideas with the mermithid *Romanomermis nielseni* in southwestern Wyoming. The species parasitizes snow pool mosquitoes, and among the five different breeding populations studied the hosts varied widely in body size distribution and parasitic load. The body size distributions of emergent adult worms varied accordingly; small adult worms emerged from small and/or crowded hosts. As indicated in Fig. 2.9(a), mean adult length varied from nearly 8 mm to almost 14 mm; among populations, the larger worms were over three times the average mass of the smaller. Figure 2.10(a) shows the complete body size distributions at two localities. As indicated in Fig. 2.10(b), the size threshold (τ is the body size when 50 per cent of the individuals are male) shifted in response to the altered adult body size distribution: the larger the worms, the larger the size threshold τ. Indeed, for the two populations, the curves of sex ratio versus adult body size (Fig. 2.10(b)) are quite different. As theoretically expected, the location with generally larger worms has the right-shifted sex ratio curve. Summarizing the position of the body size distribution with the mean adult body size $\overline{x}$ for the five study locations, Fig. 2.9(a) shows a highly significant positive relation between τ and $\overline{x}$. At emergence, nearly 80 per cent of the individuals in the populations are male (ranging from 0.89 to 0.70). This level may decrease slightly with increasing $\overline{x}$ (Fig. 2.9(b)); the correlation of -0.78 is not significant with only five

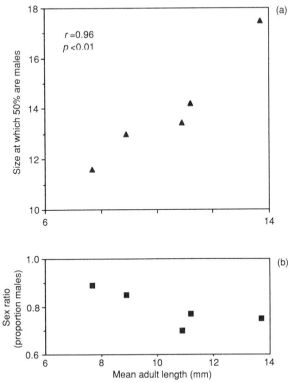

Fig. 2.9 Effects of population body size distribution on the sex ratio of *Romanomermis*. The mean length of adult nematodes is used as an indicator of the position of the body size distribution for each of five populations studied. (a) The body size τ at which the sex ratio shifts from male biased to female biased correlates positively with the position of the population body size distribution. The significance level is 0.01, even with only five points (one-tailed test). (b) The population sex ratio at emergence may show a small decline with mean adult size, but it is always quite male biased. See text for further discussion. From Blackmore and Charnov 1989.

data points, but the effect may still be real. The sex ratio is thus fairly invariant to changes in the distribution of adult body sizes.

2.6 Summary

The Fisher inheritance symmetry underlies all sex allocation theory in the sense that non-equal allocations generally reflect either its violation and/or its presence along with some asymmetry in the gains through

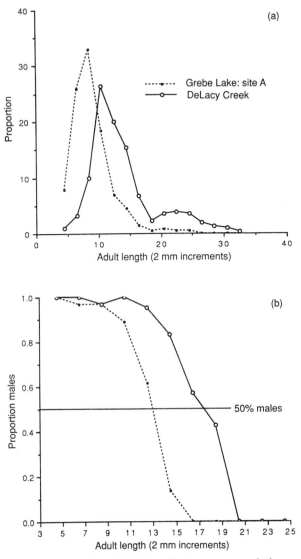

Fig. 2.10 (a) Distribution of adult body size in two populations of *Romano-mermis*. Although these particular populations are located only 30 km apart, differences in composition of the host communities and worm burden per host at these localities result in quite different body size distributions of adult nematodes. (b) Within each population, the sex ratio shifts from all male to all female as the body size increases; however, the curve of sex ratio versus adult size is itself shifted upwards in the population with mostly larger worms. The size τ at which the sex ratio is 1:1 is also indicated. Interpopulation body size differences result in different values of τ. From Blackmore and Charnov 1989.

male versus female function. As noted at the beginning of the chapter, the transformation is to switch the male and female labels. We obtain a symmetric outcome if the gain relation does not change, and an asymmetric outcome if it does. For example, simultaneous hermaphroditism has a gain asymmetry if $n_1 \neq n_2$. This chapter has discussed mostly asymmetric male/female gains. The hermaphrodite discussion was completely theoretical, whereas for sex change and environmental sex determination ESS results were derived and compared with the data. Equations (2.5) and (2.8) show the usefulness of a symmetry approach to sex allocation; the ESS sex ratios are biased, with the deviations from equality being an exact reflection of the asymmetry present in male versus female gains. Many other ESS results for sex allocations can also be written in similar symmetric (asymmetric) forms.

3

Alternative life histories, mostly about males

3.1 Introduction

The problem of sex allocation, the evolution of male-to-female ratios, is the simplest and most universal example of alternative life histories —simplest because the frequency-dependent natural selection is literally built into the problem through the fact that everyone has one mother and one father. This chapter looks at another class of alternative life histories which occur within a single sex—usually, but not always, males. While the frequency dependence of fitness is not necessary to produce some of these alternatives, it will almost always be present. However, this will differ from sex allocation in that the frequency dependence is not automatic, not built in, and therefore will usually be generated by factors external to the genetic system, commonly the environment external to the individual organism. The aim of this chapter is to model the ESS for these other alternative life histories in a way analogous to sex allocation, and to stress the usefulness of the stable age distribution symmetry illustrated in Fig. 1.4.

There are a great many examples of alternative behaviour favoured by a frequency-dependent fitness advantage. Two of these are habitat selection in relation to foraging where an individual's intake depends upon the locale decisions made by competitors (Fretwell and Lucas 1970), and mating habitat decisions made by individual males (females) in relation to the dispersion of sexual competitors (Orians 1969). Much of the data and theory on these are reviewed by Maynard Smith (1982), in a series of book chapters by Parker who pioneered the field (Parker 1978, 1982, 1984a,b; Milinski and Parker 1991), and in recent texts and multi-author volumes (Barnard 1984; Krebs and Davies 1987).

Much of the work cited above refers to decisions that individuals make on a very short time-scale, and individuals may play different roles on, for example, different days. In this chapter we shall not deal with decisions where the commitment is of short duration, but look at situations where individuals commit themselves completely to alternative life history pathways comparable to the alternative of male and female. Study of these types of alternatives has been pioneered by Gross (Gross and Charnov 1980; Gross 1982, 1984, 1985, 1991*a,b*; 1993), and I rely heavily on his work. This chapter focuses on the use of a temporal symmetry, the stable age distribution in a non-growing population, to derive ESS results. Two broad classes of models are developed (symmetric and asymmetric beginnings; the ESS results are eqns. (3.5) and (3.8) respectively) which, interestingly enough, correspond well to the two systems (bluegill sunfish and salmon) most studied by Gross. Equation (3.8), the asymmetric ESS, is a new theoretical result.

3.2 Bluegill sunfish

Alternative male mating behaviour is a common and widespread phenomenon in fish (Gross 1984, Table 1). If male parental care is the primary behaviour, the alternative is usually cuckoldry (fertilization of eggs which are cared for by another male) either through sneaking or female mimicry. Both these alternative types of behaviour are found in bluegill sunfish (*Lepomis macrochirus*) and other species in the family Centrarchidae, native to eastern North America (Gross 1982, 1984, 1991*a*). Gross (1979) discovered the cuckoldry habit and was able to show that it represented a completely alternative life history (Gross and Charnov 1980). Figure 3.1 shows the two alternatives (contrasted with coho salmon) along with the monomorphic female life history. At Gross's Canadian locality, parental bluegill males mature at 7–8 years. In the spring and early summer they construct nests arranged in dense aggregations or colonies in the shallow littoral zone. The nests are shallow depressions on the lake bottom. Females arrive at the colony in schools and are courted by the parentals; the females subsequently enter the nests and spawn, releasing eggs repeatedly in small batches, each of which is fertilized by the male in attendance. Males may receive eggs from many females, and females may spawn in many nests; spawning within a single colony is highly synchronized, taking place over about a single day. After spawning the females leave the colony area while the parental males remain to care for 'their' eggs.

Cuckolder males mature at 2 years (much more rarely 1 or 3 years): when very small (aged 2–3 years) they spawn as sneakers; when older

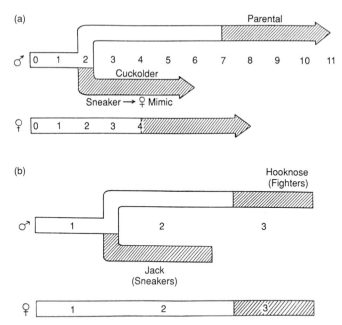

Fig. 3.1 (a) Male bluegill sunfish have two alternative life history pathways: one leading to cuckoldry and the other to parental care (see text for further details). These data are from Gross (1982) and describe a bluegill population in Lake Opinicon, Ontario. Hatched areas indicate reproductive maturity. (b) Salmon males have two alternative life history pathways: jacks and hooknoses. The data are for coho salmon and are taken from Gross (1984); the numbers indicate age in years.

(4–5 years) they operate as female mimics (or satellites) using deception to gain access to the nest to pair with a female during her spawning dips. While only about 20 per cent of a cohort matures as cuckolders (95 per cent binomial confidence interval is 11–31 per cent (Gross 1991*a*)), the large age difference between them and the parentals means that the cuckolder morph comprises most of the reproductive male population (ratio about 6:1). Gross (1991*a*) studied reproductive behaviour for several colony types with a range of water depths and cover (provided by vegetation and debris). The cover provides cuckolders safety from predators as well as places to hide from the nest-defending parental males.

Gross (1991*a*) used experimental manipulation to demonstrate negative frequency dependence for the fitness pay-offs for cuckolders; the exact results depended heavily on colony type (amount of cover etc.). The one variable unknown in his system was how the cuckolders distributed themselves among the various types of colony. However, his data allowed the following two calculations. He took the estimated cuckolder population and divided it among the colony types using one of two rules. First, they were assigned at random, with each colony type weighted by its size (number of parental males). The second rule assigned cuckolders to colonies so as to maximize their average success relative to parental males (this is the ideal free mating distribution of habitat selection theory (Fretwell and Lucas 1970)). Under the observed cuckolder-to-parental ratio of 6:1, these two habitat assignment rules mean that the cuckolders fertilize somewhere between 11 per cent (random distribution) and 23 per cent (ideal free) of the eggs in a breeding season. The fact that this proportion (0.11–0.23) is near the proportion (0.11–0.31) of 2 year olds who become cuckolders will be interpreted later in this chapter.

3.3 Salmon

Many species in the fish family Salmonidae (trout, salmon, charr, white-fish) show alternative male life histories (Gross 1984, 1985, 1991*b*; Maekawa and Hino 1987). Gross (1984, 1985) has made a particularly elegant study of coho salmon (*Oncorhynchus kisutch*) from the west coast of North America. The alternative life histories are shown in Fig. 3.1. The species is semelparous. At age 2 years from hatching, small cryptically coloured males (called 'jacks') leave the ocean and breed in freshwater streams as sneakers. At age 3 years the much larger and brightly coloured hooknose males leave the ocean to breed. The females also breed at age 3 years, excavating a gravel nest in the stream bottom. All fish die after spawning, although females may defend their nest for a brief period. Proximity to the female at the moment of egg laying is the key to fertilization success. Hooknose males use fighting to gain proximity while jacks use sneaking, making use of refuges (rocks etc.) to escape aggression from the larger males. Gross (1985) showed that males intermediate in size between the jacks and the hook-noses were at a severe disadvantage as they could neither successfully fight nor sneak and hide. By making some plausible assumptions about fertilization success as a function of proximity to the female, Gross (1985) estimated that a jack's fertilization success was about 66 per cent that of a hooknose on a per female basis. A jack's breeding life-span (time on the breeding grounds) was also about two-thirds that of a hooknose. Thus each jack on the breeding grounds had about 44 per cent ($2/3 \times 2/3$) of the reproductive success of each hooknose. However, jacks mature a year earlier (Fig. 3.1) and have an ocean survivorship to maturity more than double that of the hooknoses (0.13 versus 0.06). Thus each jack has a life-time reproductive success *approximately* equal to that of each hooknose ($0.44 \times 0.13/0.06 = 0.95$), at least in this population (there are no standard errors available for these estimates).

Gross (1993) admits that this calculation is not particularly precise and indeed suggests that the number (here 0.95) ought to be above unity, representing an equilibrium which does not equalize the average life-time fitness pay-offs for the two pathways. His suggestion is based on a growing body of evidence that precocious maturity as a jack is linked to large size as a fry (Bilton 1980; Borghetti *et al.* 1989; Rowe and Thorpe 1990); faster-growing individuals apparently choose the jack life history (Fig. 3.2). The difference between the sunfish and the salmon lies in the apparent role of juvenile growth in the decision to adopt the life-style of precocious maturation. Enhancement of juvenile growth results in more individuals becoming jacks in the salmon while a similar feeding and growth enhancement experiment shows no effect for bluegill (Gross and Philipp, unpublished; quoted by Gross 1993). Gross (1993) notes that this growth effect should follow if larger juveniles were relatively better, in a fitness sense, at being jacks. From this he correctly concludes (see below) that at the ESS frequency a jack ought to achieve greater life-time fitness compared with a hooknose.

In the next two sections ESS models are developed for bluegill sunfish and salmon respectively.

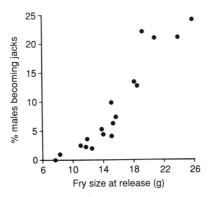

Fig. 3.2 Relationship between the size of coho salmon fry released into the ocean and the percentage of adult males that return to the hatchery as jacks. Data are from an experimental study by Bilton (1980). Figure reproduced from Gross (1991*b*).

3.4 ESS theory: symmetric beginnings

Suppose that we have two alternative life histories, as found in bluegill sunfish (Fig. 3.1); at or before age 2, when the alternatives first show themselves, all individuals are equivalent with respect to their reproductive possibilities down either of the pathways. They may differ in size etc., but the differences do not translate to differential (asymmetric) reproductive opportunities down one versus the other pathway. For this situation a wide variety of genetic systems will result in an ESS proportion q at the time of division which equalizes the average fitness gains for the two life histories (Slatkin 1978; Maynard Smith 1982). While it is possible to test this by actually observing life-time reproductive success (Clutton-Brock 1988), the immense difficulty of such studies suggests adopting an alternative research strategy; the temporal symmetry provided by a non-growing population combined with a stable age distribution (a stationary population) underlies one useful alternative (Gross and Charnov 1980).

Consider the following for the bluegill sunfish example. Let $L(x)$ be the probability that an individual is alive at age x, and $b(x)$ be the number of offspring that an individual will father (eggs it will fertilize) at age x. Because $b(x)$ and $L(x)$ will differ for each pathway, let p and c designate parental and cuckolder respectively. The population is assumed to be stationary. If the fitness is to be the same over each pathway (at the ESS), it follows that

$$\int_{x=2}^{\infty} L_c(x) b_c(x)\, \mathrm{d}x \ = \ \int_{x=2}^{\infty} L_p(x) b_p(x)\, \mathrm{d}x \quad . \tag{3.1}$$

Each integral is over an individual's life-time for each pathway. I now show that a more easily testable life history prediction can be derived from the relationships in eqn. (3.1). Let n males reach age 2 each year. If a proportion q of these go down the sneaker-satellite (cuckolder) pathway, then the number of age x c males is $qnL_c(x)$. Because each of these fertilizes $b_c(x)$ eggs, the total number of eggs fertilized by all c males each year is

$$qn \int_{x=2}^{\infty} L_c(x) b_c(x)\, \mathrm{d}x \quad . \tag{3.2}$$

The same function for the parental males is

$$(1 - q)n \int_{x=2}^{\infty} L_p(x) b_p(x) \, dx \quad . \tag{3.3}$$

The ratio of eqn. (3.2) to eqn. (3.3) is the ratio of the eggs fertilized by cuckolder males to the eggs fertilized by the parental males *in each breeding season*. Let H be the proportion of eggs in a breeding season fertilized by c males. Recall the conditions for an ESS q (eqn. (3.1)). Forming the ratio of eqn. (3.2) to eqn. (3.3), we have

$$\frac{qn \int_{x=2}^{\infty} L_c(x) b_c(x) \, dx}{(1 - q)n \int_{x=2}^{\infty} L_p(x) b_p(x) \, dx} = \frac{H}{1 - H} \tag{3.4}$$

or

$$q = H \quad . \tag{3.5}$$

That is, the equilibrium (ESS) proportion of males at age 2 entering the sneaker-satellite (cuckolder) pathway should be equal to the proportion of eggs fertilized in each breeding season by all the sneaker-satellite males. We have thus transformed the life history calculations into a set of much easier short-term observations (i.e. estimating H). All this follows from the assumption of a stationary population, from the equivalence of the $L(x)$ schedule to the age distribution (Fig. 1.4).

We can now see the logic behind Gross's (1991*a*) calculation of the proportion of eggs fertilized in a single season by all cuckolder bluegill males. This number is H, and by the theory developed here it should be equal in ESS to the proportion of the males who enter the cuckolder pathway at age 2 years; the equivalence is approximately true for bluegill sunfish. The hypothesis that $q \approx H$ in ESS was first applied to bluegills by Gross and Charnov (1980). A discrete-generation version of this prediction was derived and tested for fig wasps by Hamilton (1979) (reviewed by Maynard Smith 1982). The discrete-generation model does not require the assumption of a stable age distribution.

3.5 ESS theory: asymmetric beginnings

Consider a situation like that found in salmon (Figs. 3.1 and 3.2) where some condition factor like juvenile size translates (or can translate) differentially (asymmetrically) into reproductive ability down the two pathways. Here I give two derivations for ESS results: the one provided here in the text is meant to be intuitive, whereas in the Appendix to the chapter I use a more formal ESS argument by examining the fitness fate of a rare mutant. Of course, the ESS answers are the same. Let P stand for the primary pathway (hooknose) and A for the alternative (jack). The discrete-generation life history is modelled here (the salmon), but the formalism is also correctly interpreted for over-lapping generations, just like eqn. (3.5). H is defined as the proportion of the eggs that are fertilized by A males each year (H may change as a function of the number of individuals who are A, and the calculations given below are for H in the neighbourhood of the ESS decision), z as the condition variable like juvenile size or juvenile growth rate, and $\omega_A(z)$ and $\omega_P(z)$ as the relative fitnesses of an individual of condition z if it is A or P respectively ($\overline{\omega}_A$ and $\overline{\omega}_P$ are the average relative fitness of those playing A and P); there are N_A A males and N_P P males at the time that the choice of pathway is made. I make two further assumptions as illustrated in Fig. 3.3. First, ω_A/ω_P increases with z; it is always relatively better to be A (a jack) at larger z (in better condition). This is an asymmetry in the way condition z maps to relative fitness for the two pathways. If ω_A and ω_P change with z

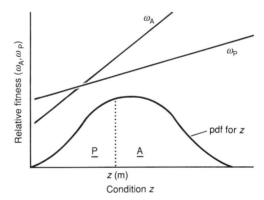

Fig. 3.3 ESS under asymmetric beginnings. Some externally imposed condition factor z like juvenile size translates differently (asymmetrically) to relative fitness down the two pathways ω_A and ω_P, where P is the primary pathway and A is the alternative pathway. Here an individual at larger z is relatively better at being A; ω_A/ω_P increases with z. The population's probability distribution function for z at the age that the choice of pathways is made is denoted by pdf. At some intermediate z, here labelled $z(m)$, an individual is indifferent (in a fitness sense) to being A or P. This *equalization of marginal fitness* can be used to derive ESS results, as shown in Fig. 3.4.

in the same way (i.e. ω_A/ω_P does not alter with z), there is no asymmetry and natural selection does not favour using z to determine the choice of pathway. This is essentially the symmetric case modelled for bluegill sunfish (eqn. (3.5)). The second assumption is that the distribution of conditions (the population distribution function (pdf) for z) is externally imposed; a juvenile individual has a certain chance of being some z and can only choose to be A or P as a function of its condition. Of course, the ESS is for larger individuals to play A, smaller ones P (Charnov *et al.* 1978; Parker 1982). At some intermediate z value, labelled $z(m)$ in Fig. 3.3, an individual's fitness is the same regardless of whether it plays A or P; $z(m)$ is the condition that equalizes the marginal fitness for the two pathways.

This ESS rule can be used to derive formulae for the ESS proportion of juveniles who adopt one rather than the other pathway. In the calculation provided here we make the implicit assumption that mutants who alter their $z(m)$ do not alter the value of H, which is dependent upon the background population of males. In symbols (z is $z(m)$) the ESS rule is

$$\frac{\omega_A(z)}{N_A \overline{\omega}_A} H = \frac{\omega_P(z)}{N_P \overline{\omega}_P} (1 - H) . \tag{3.6}$$

Since $\omega_A(z)$ and $\omega_P(z)$ are defined in a purely relative way, nothing in this rule is changed if we multiply the ω_A and $\overline{\omega}_A$ by any positive number (and similarly for ω_P and $\overline{\omega}_P$). As shown in Fig. 3.4, we adopt the convention that $\omega_A = \omega_P$ at $z(m)$. The rule now reduces to

$$\frac{N_A}{N_P} = \frac{H}{1 - H} \frac{\overline{\omega}_P}{\overline{\omega}_A} . \tag{3.7}$$

To simplify this for comparison with the bluegill sunfish, the symmetric answer (eqn. (3.5), $q = H$), we note that $N_A/N_P = q/(1 - q)$ and define $\overline{\omega}_A/\overline{\omega}_P = 1 + \beta$. Equation (3.7) now reduces to

$$q = \frac{H}{1 + \beta(1 - H)} \tag{3.8}$$

Therefore, q equals H multiplied by a factor which incorporates asymmetry for the impact of juvenile condition on relative fitness for the two pathways. H/q is the average total fitness (actual eggs fertilized) by an individual playing A (jack) and equals $1 + \beta(1 - H)$ which is greater than unity, as also argued by Gross (1993). Figure 3.5 illustrates how much q deviates from H for several β values; it requires rather large β values (greater than $1/2$) for q to be much larger than H at the ESS.

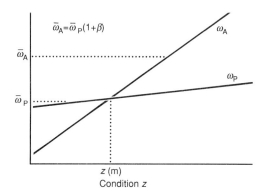

Fig. 3.4 ESS under asymmetric beginnings. Following the text (or Appendix) derivation I adopt the convention that $\omega_A = \omega_P$ at z(m); this is allowed since the ω curves give only relative fitness as a function of z within each pathway. If $\overline{\omega}_A/\overline{\omega}_P = 1 + \beta$, the ESS proportion q of juveniles who go down the A (jack) pathway is $H/[1 + \beta(1 - H)]$ where H is the proportion of females mated to A males during each breeding season.

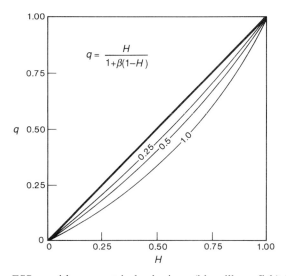

Fig. 3.5 The ESS q with symmetric beginnings (bluegill sunfish) is for $q = H$; this figure shows how far the ESS q is pushed away from H by the condition dependence of relative fitness ($\beta > 0$) for three β values (0.25, 0.50, and 1.0).

The reason for developing alternative life histories as done here is to set up an exact correspondence to sex allocation (Chapter 2). With sex allocation H always equals $1/2$; in this more general situation H changes with q against the individual's background of reproductive opportunities. If we set $H = 1/2$ in eqn. (3.7), we obtain $q/(1 - q) = \overline{\omega}_P/\overline{\omega}_A$, the same as the sex allocation equations (2.5) and (2.8).

3.6 One non-intuitive prediction

According to eqn. (3.8), β controls the extent of deviation from the symmetric ESS of $q = H$. Consider Fig. 3.6 which shows two extreme cases. Both share a common $\omega_A(z)$ curve. ω_{P_2} is a steep curve, similar to the ω_A curve, but still having ω_A/ω_P increasing with z so as to favour a conditional decision about the pathway chosen. ω_{P_1} is a flat curve showing no effect of z on ω_P; this case also favours a conditional decision. The question is: which of these two cases gives the largest β, the largest deviation from $q = H$? At first glance we might guess the ω_{P_1} case since here the ω_A and ω_P curves are the most different, the most asymmetric. However, the answer is that β will generally be larger in the ω_{P_2} case, when the ω_P and ω_A curves shows greater similarities (yet are still different enough to favour a conditional decision). Recall that $1 + \beta$ is the ratio of the average ω_A to the average ω_P at the ESS; provided that both curves increase with z, β will be larger when both curves are steep. This prediction violates my intuition. It appears to be true for sex-changing fish (Charnov 1982, p. 139; eqn. (2.5) here). Under protandry the male curve is flat while the female curve increases with

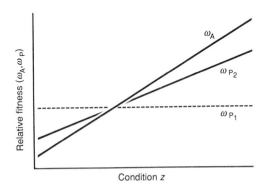

Fig. 3.6 β is larger if both the ω_A and ω_P curves are steeply increasing. It is larger for the pair ω_A, ω_{P_2} compared with ω_A, ω_{P_1}. Thus similar (increasing) ω curves favour the greater deviation from the symmetric ESS of $q = H$.

age (size); under protogyny both curves go up. Protogyny has the more biased breeding sex ratios (i.e. deviation from $q = H = 1/2$).

3.7 Summary

This chapter uses symmetry in two ways. The first, as shown by the bluegill sunfish theory leading to eqn. (3.5) ($q = H$ in ESS), is the usefulness of the assumption of a stationary population. For this situation we need not measure life-time fitness for each of the alternative pathways and can substitute a theoretical prediction about the eggs fertilized at a single point in time by all the members of a pathway. It should be much easier to estimate H than to estimate life-time fitness.

The second use of symmetry is to ask if some juvenile condition variable, such as size or growth rate, translates differentially (asymmetrically) into reproductive ability down the two pathways. If ω_A/ω_P does change with condition z then the decision to adopt one or the other alternative ought to be dependent upon condition, resulting in an ESS which can be expressed in terms of the average ω values within each pathway (eqn. (3.8)). If ω_P/ω_A does not change with z, the curves are symmetric and an individual ought to ignore its condition in its choice of pathway. If z is not used in the choice of pathway, $\overline{\omega}_A$ will equal $\overline{\omega}_P$, and the ESS is again for $q = H$ (formal derivation in the Appendix). I have illustrated these two cases with the bluegill sunfish and salmon examples. The salmon give clear evidence for a condition dependent (growth-dependent) decision to become a hooknose versus a jack. Condition dependence has not been shown for the bluegill choice of cuckolder versus parental. Condition dependence is probably the more common form of decision in nature.

3.8 Appendix

This Appendix develops a more formal ESS approach to the 'asymmetric beginnings' case (i.e. leading to eqns. (3.7) and (3.8)). The following definitions are made for discrete generations: H is the proportion of eggs fertilized by A males; z is a condition variable such as juvenile size or growth rate; $g(z)$ is the probability density for z among the juveniles; $\omega_A(z)$ ($\omega_P(z)$) is the relative fitness of an individual of condition z who reproduces as an A (P); $r(z)$ ($\tilde{r}(z)$) is the proportion of males of condition z who reproduce as A for the wild (mutant) type; E is the number of eggs available to be fertilized; N is the number of males present in the population (a large number).

The total fitness W of a rare mutant is the sum of the eggs it will fertilize through being A plus those through being P. The mutant is only allowed to alter its response $\tilde{r}(z)$ to condition z. We have

$$W = \frac{\int_0^\infty g(z)\tilde{r}(z)\omega_A(z)\,dz}{N\int_0^\infty g(z)r(z)\omega_A(z)\,dz} HE + \frac{\int_0^\infty g(z)[1 - \tilde{r}(z)]\omega_P(z)\,dz}{N\int_0^\infty g(z)[1 - r(z)]\omega_P(z)\,dz}(1 - H)E$$

We wish to find the ESS $r(z)$, which is a function such that no mutant altering its $r(z)$ (to $\tilde{r}(z)$) can increase its fitness W. We do this by looking at the signs of $\partial W/\partial \tilde{r}(z)$ for all values (i.e. small intervals of z) of z. For some specified z, the sign $(+, 0, -)$ of this derivative is the same as the sign of

$$\frac{\omega_A(z)H}{D_1} - \frac{\omega_P(z)(1 - H)}{D_2} \qquad (3.9)$$

where

$$D_1 = \int_0^\infty g(z)r(z)\omega_A(z)\,dz$$

$$D_2 = \int_0^\infty g(z)[1 - r(z)]\omega_P(z)\,dz \quad .$$

We consider two cases: first where ω_A/ω_P increases with increasing z, and second where ω_A/ω_P is always the same. In the first case, condition z maps *asymmetrically* to relative fitness ω_A/ω_P; in the second it does not.

CASE 1 ω_A/ω_P goes up with z.
The sign of eqn. (3.9), for any z, is the same as the sign of

$$\frac{\omega_A(z)}{\omega_P(z)}\frac{H}{D_1} - \frac{1 - H}{D_2} \qquad (3.10)$$

For small z (small ω_A/ω_P) this will be less than zero; for large z (large ω_A/ω_P) it will be greater than zero. At some intermediate value of z, it will equal zero. Thus the ESS $r(z)$ is for small z to be P and large z to be A; the changeover point $z(m)$ is where eqn. (3.10) equals zero or

$$\frac{\omega_A}{\omega_P} \frac{H}{1-H} = \frac{D_1}{D_2} \tag{3.11}$$

where ω_A and ω_P are at $z(m)$ and D_1 and D_2 now have the forms

$$D_1 = \int_{z(m)}^{\infty} g(z)\omega_A(z)\,dz \tag{3.12}$$

$$D_2 = \int_{0}^{z(m)} g(z)\omega_P(z)\,dz \tag{3.13}$$

We can rewrite (3.12) as

$$D_1 = \left[\int_{z(m)}^{\infty} g(z)\,dz\right] \frac{\int_{z(m)}^{\infty} g(z)\omega_A(z)\,dz}{\int_{z(m)}^{\infty} g(z)\,dz}$$

The term on the extreme right is the average ω_A denoted here as $\overline{\omega}_A$. D_2 may be similarly rewritten so that eqn. (3.11) becomes

$$\frac{\omega_A}{\omega_P} \frac{H}{1-H} = \frac{\left[\int_{z(m)}^{\infty} g(z)\,dz\right]\overline{\omega}_A}{\left[\int_{0}^{z(m)} g(z)\,dz\right]\overline{\omega}_P} \tag{3.14}$$

Finally, the ratio $\left[\int_{z(m)}^{\infty} g(z)\,dz\right] \bigg/ \left[\int_{0}^{z(m)} g(z)\,dz\right]$ is the ratio of the number of A males to the number of P males, denoted N_A/N_P in the chapter. Equation (3.14) can thus be written as

$$\frac{\omega_A}{\omega_P} \frac{H}{1-H} = \frac{N_A \overline{\omega}_A}{N_P \overline{\omega}_P} \tag{3.15}$$

This is the same ESS result as eqn. (3.6).

CASE 2 ω_A/ω_P does not change with z.
The derivatives $\partial W/\partial \tilde{r}(z)$ have the same sign(s) as

$$\frac{\omega_A(z)}{\omega_P(z)} \frac{H}{D_1} - \frac{1-H}{D_2} \tag{3.16}$$

However, since ω_A/ω_P is the same for each and every z value, all the derivatives must have the same sign; thus they must each be equal to zero in ESS. Recall that since $\omega_A(z)$ and $\omega_P(z)$ are measured only in a relative way, we can simply agree to view the two curves as falling on top of one another, i.e. $\omega_A(z) = \omega_P(z)$. $r(z)$ must now be independent of z, so that with $\omega_A(z) = \omega_P(z)$, eqn. (3.16) set equal to zero yields

$$\frac{H}{1-H} = \frac{D_1}{D_2} = \frac{\displaystyle\int_0^\infty g(z)r\omega_A(z)\,\mathrm{d}z}{\displaystyle\int_0^\infty g(z)(1-r)\omega_P(z)\,\mathrm{d}z} = \frac{r}{1-r}$$

or $r = H$ in ESS (the same as eqn. (3.5); q is the same as r).
   If $\omega_A(z)/\omega_P(z)$ does not change with z, the curves are symmetric, not different, and this condition z is not to be used in the choice of pathway.

4

Indeterminate growth

4.1 Introduction

Although many species of animals, including a diverse array of birds, mammals, cephalopods, and terrestrial arthropods, cease growing after they reach sexual maturity, this pattern of determinate growth is the exception rather than the rule. Most other vertebrates and invertebrates show indeterminate growth where they begin reproducing before they attain maximum body size. Prior to maturation, energy is allocated to maintenance and growth, whereas after maturation it is also allocated to reproduction. Indeterminate growth is commonly described by expressions such as the Bertalanffy equation; Fig. 4.1 shows a typical life history where the rate of growth declines with age and the animal eventually approaches some asymptotic length if it survives for long enough. At the age at maturation α the individual is typically 50–80 per cent of the asymptotic length (15–50 per cent of the weight) depending upon the species.

The Bertalanffy equation is the most commonly used descriptor of growth (Fig. 4.1), and its two-parameter form is usually written as $\ell_x = \ell_\infty [1 - \exp(-kx)]$ where ℓ_x is length and x is age. ℓ_∞ is the asymptotic length and k is the growth coefficient, which has units of time^{-1} and indicates how fast the asymptote is approached. This equation passes through the origin ($\ell_0 = 0$), but a more general three-parameter form discussed later in this chapter relaxes this assumption.

Thirty years ago Beverton and Holt (1959) and Beverton (1963) pioneered the comparative study of fish (indeterminate growth) life histories by showing that, within limited taxonomic boundaries (such as within the cod or herring family), there existed certain patterns in growth and mortality across species (or populations within a species). These patterns relate to the values of certain dimensionless numbers made up from components of the life history as defined by growth,

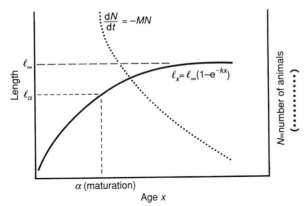

Fig. 4.1 A life history under indeterminate growth. Growth in length ℓ_x follows the Bertalanffy equation $\ell_x = \ell_\infty [1 - \exp(-kx)]$; mass follows length to a power near 3. Maturation is at age α and length ℓ_α, and ℓ_∞ is the asymptotic length; thus ℓ_α/ℓ_∞ is the relative size at maturity. The adult instantaneous mortality rate is M. In practice M is an average over the adult life-span weighted towards the younger adults. Beverton (1963) developed and used a statistic equivalent to calculating the average M for the stable age vector.

mortality, and maturation. Three patterns are reviewed by Cushing (1968), Pauly (1980), and most recently Beverton (1992), and relate to the life history illustrated in Fig. 4.1. First, within each group the adult instantaneous mortality rate M and the Bertalanffy growth coefficient k are positively related to each other so that the ratio M/k tends to be relatively constant. The second pattern is that the length at maturity ℓ_α is positively related to the Bertalanffy asymptotic length ℓ_∞ so that the relative length at maturity ℓ_α/ℓ_∞ tends to be a constant value within a taxon (e.g. *c.* 0.75 for clupeids and *c.* 0.55 for cod).

Charnov and Berrigan (1991*a*) noted that the dimensionless number αM has an interesting relation to k/M and ℓ_α/ℓ_∞. Suppose that growth follows the two-parameter Bertalanffy equation; then we have

$$\frac{\ell_\alpha}{\ell_\alpha} = 1 - \exp(-k\alpha) = 1 - \exp\left(-\frac{k}{M}\alpha M\right) \qquad (4.1)$$

This means that species with the same ℓ_α/ℓ_∞ and k/M will have the same αM; for such species the adult instantaneous mortality rate M will be inversely proportional to the age at maturity α, with the constant of proportionality equal to $-(M/k)\log_e (1 - \ell_\alpha/\ell_\infty)$.

The third pattern noted by Beverton and Holt (1959) relates to the growth parameters ℓ_∞ and k. Within a closely related group (e.g. a genus) they showed a negative relation: the faster the animal grows, the smaller it is as an old adult. Of course, within a single data set on growth the two parameters will have a negative covariance, since they are estimated from the same data. However, the negative relationship observed and discussed here is much too large to be due to this statistical artefact. These $k-\ell_\infty$ relations tend to be linear on a log–log plot, indicating that the separate growth curves (among the species, populations, etc.) are linked by the relation $\ell_\infty = Dk^{-h}$ (Pauly 1980; Munro and Pauly 1983; Pauly and Munro 1984; Moreau *et al.* 1986, reviewed by Longhurst and Pauly 1987, Ch. 9). Pauly noted that the dimensionless slopes h cluster around 0.5. For example, such a plot for 100 closely related (several species) Tilapia populations gives a linear relation with a functional regression (Ricker 1973, 1975) estimate of h of 0.57 (Moreau *et al.* 1986).

The approximate invariance within certain taxonomic boundaries for the dimensionless numbers ℓ_α/ℓ_∞, M/k, and h will be collectively labelled the Beverton–Holt invariants. In this chapter we review evidence for them within several animal groups (fish, pandalid shrimp, sea urchins, snakes, and lizards) and then use life history theory to ask what they can tell us about symmetries at the level of life history trade-offs. To preview the conclusions: M/k and ℓ_α/ℓ_∞ will be argued as invariants representing the outcome of symmetries in two particular life history constraints. The $k-\ell_\infty$ relation will end up indicating a growth rate trade-off; we shall examine when a negative relation is expected and why it contains useful information about this particular trade-off. Discussion of the $\ell_\infty-k$ relations will be deferred until the end of the chapter. Two cases where ℓ_α/ℓ_∞ and M/k are not at all invariant across populations within a species will also be discussed.

The statistics, including 95 per cent confidence intervals and/or standard errors, are relegated to the figure (table) captions. Most fitted equations are ordinary regressions since I believe that the x variable is usually known with much greater precision than the y variable; when this is clearly not true (e.g. $k-\ell_\infty$ relations), I have used functional regressions (major axis) which place equal error variance on each axis (Ricker 1973, 1975; Harvey and Pagel 1991). Sometimes when proportionality is obvious I have estimated the slope using a one-parameter regression through the origin. Proportionality is usually tested by finding a slope of ± 1 on a log–log plot. I consider a proportional relationship to indicate approximate invariance if the correlation is reasonably high (say, above 0.75).

To summarize briefly, in this section we shall review evidence for the invariance of the dimensionless numbers M/k (or k/M), αM, and

ℓ_α/ℓ_∞. Later we shall examine h in the (proposed) trade-off relation $\ell_\infty = Dk^{-h}$.

4.2 Fish

Clupeomorpha

Beverton (1963) provides a comprehensive review of the M/k and ℓ_α/ℓ_∞ numbers for many populations within several species of clupeomorph fishes in the families Engraulidae and Clupeidae—the herring, anchovies, and sardines. The data strongly support the approximate invariance of these numbers between populations within a species. While he documents some differences between species (e.g. $0.68 < \ell_\alpha/\ell_\infty < 0.81$), the species are similar enough to justify pooling the data for this overview discussion (species which are very long lived have M/k and αM numbers lower than the averages discussed here (Beverton 1992)). Figure 4.2(a) shows a plot of $1/T_{\max}$ versus k, where $T_{\max}$ is the age of the oldest

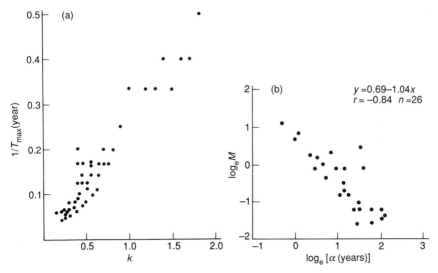

Fig. 4.2 (a) Growth coefficient k versus $1/T_{\max}$, where $T_{\max}$ is the maximum life-span, for 45 populations of 10 species of fish in the families Engraulidae and Clupeidae. Because the adult instantaneous mortality rate M is proportional to $1/T_{\max}$, $M = 1.5k$ for these fish. Data from Beverton (1963). (b) The adult instantaneous mortality rate M is inversely proportional to the age of maturity α for fish in the families Clupeidae and Engraulidae. Data are from Beverton (1963) and only include populations with direct estimates of M (which is why there are fewer data points than in (a)). (Standard errors: slope, 0.14; intercept, 0.17).

individual observed in a large sample. Beverton (1963) showed that T_{max} is a well-behaved statistic (for large samples, say more than 5000) which for these fish is highly correlated with the adult mortality rate M, so that $M = g/T_{max}$ with $g \approx 6$. (Beverton had $g = 6.4$ on average.) In a much larger and taxonomically diverse sample of animal species, Hoenig (1983) confirmed Beverton's relation, with a very similar g value (he obtained $g = 5.5$ for fish by a functional regression). Application of this relation (with $g = 6$) to Fig. 4.2(a) where $kT_{max} = 4$, yields $M = 1.5k$ for the Clupeomorpha. Figure 4.2(b) shows a smaller sample of populations where M is known and indicates that M and α are inversely proportional with $M\alpha = 2$. These two invariants give $\ell_\alpha/\ell_\infty = 0.74$, right in the centre of the Clupeomorpha range. Beverton (1992) re-estimated g for a better-known, although smaller, sample of long-lived clupeids, and this time obtained $g \approx 3$. It is difficult to reconcile this new value of g with previous estimates; see Beverton (1992) for a detailed discussion of some of these issues, as well as consideration of cod and flatfish.

M/k for fish in general

Pauly (1980) compiled data on M and k for 175 fish stocks (separate breeding populations) representing 110 species from all over the world. Several fish families are represented by 10–17 stocks each, and the within-family regressions are generally significant (0.05 level or better) with positive slope. In this data set the family Scombridae (tuna and mackerel) has the largest sample size (17) and shows the best within-family M–k relation; this is plotted in Fig. 4.3 and gives $M = 1.7k$. The Pauly compilation did not show a positive relation for the family Lutjanidae (snappers) probably because of the small number of populations (seven) in the sample; a much larger data set compiled by Ralston (1986) shows a highly significant relation with $M = 2k$ (Fig. 4.4).

Figure 4.5 shows a plot of $\log_e M$ versus $\log_e k$ for all 175 stocks using ordinary regression. The relation is linear with unit slope, making M/k a constant equal to 1.65. Various alternative statistical assumptions (or estimation procedures) which could put M/k as high as 2.1 are discussed in the figure caption, but these simply establish $0.48 < k/M < 0.63$ for fish overall which is a fairly narrow range. An alternative method of analysis might be to fit $\log_e M$ versus $\log_e k$ using, say, family averages. This has been done for the 12 families with more than five stocks each and the fitted regression is $\log_e M = 0.55 + 0.91 \log_e k$ (with $r = 0.88$), not very different from Fig. 4.5. Later we shall consider what life history theory might say about this narrow k/M range.

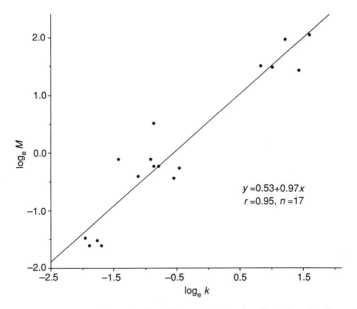

Fig. 4.3 $\text{Log}_e M$ versus $\log_e k$ for 17 stocks in the fish family Scombridae (tuna and mackerel). The slope of unity means that M/k is a constant equal to 1.70. (Standard errors: slope, 0.08; intercept, 0.11). Data compiled by Pauly (1980).

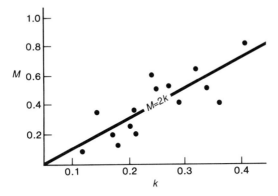

Fig. 4.4 M versus k for 15 stocks in the fish family Lutjanidae (snappers). Data from Ralston (1986), compiled by Longhurst and Pauly (1987).

With a 'typical fish' having $\ell_\alpha/\ell_\infty \approx 0.65$ and $1.65 < M/k < 2.1$, by eqn. (4.1) αM is in the approximate range 1.75–2.2. Fish with smaller ℓ_α/ℓ_∞ and smaller M/k values will, of course, have smaller αM values (Beverton 1992).

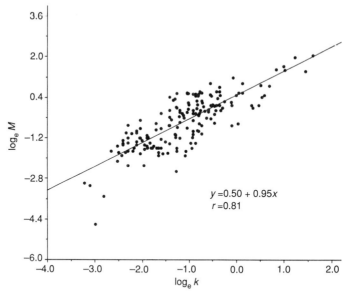

Fig. 4.5 $\log_e M$ versus $\log_e k$ for 175 fish stocks representing 110 species. The slope near unity makes M/k a constant equal to 1.65. (Standard errors: slope, 0.05; intercept, 0.075). There are alternative ways of estimating the M/k ratio. The regression $M = Ck$ has a 95 per cent confidence interval for C of 1.48–1.71 (mean, 1.60); the regression $k = (1/C)M$ has a confidence interval of 1.75–2 (mean, 1.89). The same one-parameter models for the means for the 12 families with more than five stocks each produces confidence intervals of 1.43–2.0 (mean, 1.72) and 1.59–2.2 (mean, 1.82) respectively. Finally, the simple average for the 175 M/k ratios is 2.07 (standard error, 0.1) and the geometric mean is 1.75. On the basis of these alternative estimation schemes $1.6 < M/k < 2.1$ and $0.48 < k/M < 0.63$. Pauly (1980) claimed a rather more complex relation between $\log_e M$, $\log_e k$, and the mean environmental temperature for the stock. He fitted a multiple linear regression of $\log_e M$ on $\log_e k$ and $\log_e$ temperature. However, a regression of $\log_e$ temperature on the residuals of this figure shows a very weak positive relation ($r^2 = 0.07$), while a $\log_e M/k$, $\log_e$ temperature regression is also very weak ($r^2 = 0.09$). Thus, contrary to Pauly, I suggest that the evidence for a temperature effect is quite weak. Data compiled by Pauly (1980).

Walleye and brown trout: violation of the Beverton–Holt invariance

Between-population comparisons within two species with large sample sizes do not show M/k and ℓ_α/ℓ_∞ invariance.

Beverton (1987) analysed data compiled by Colby *et al.* (1979) and Colby and Nepszy (1981) on the geographical variation of life history for 13 separate populations of the North American pike perch or walleye

(*Stizostedion vitreum*). The data, summarized in Table 4.1, are particularly interesting as the populations studied ranged in latitude from Ontario, Canada, in the north to Texas, USA, in the south. T_{max} ranged from 20 years (Ontario) down to 3 years (Texas). All variables are closely related to temperature over the growing season. It should be noted that this species does not show invariance in either M/k or ℓ_α/ℓ_∞ over this geographical range. For example, kT_{max} ranges from 1.0 in Ontario to 3–4 in Texas and California; ℓ_α/ℓ_∞ shows a similar alteration, ranging from 0.5 in Ontario up to 0.9 in Texas. However, as shown in Fig. 4.6, αM is fairly constant for this life history and equals 2.30 (if $g = 6$), not very different from the 'typical fish' values (1.75–2.2). Equation (4.1) shows how αM can remain approximately invariant if both k/M and ℓ_α/ℓ_∞ increase; this appears to happen for the walleye in North America.

Vollestad *et al.* (1993), in a comparison of 29 populations of brown trout (*Salmo trutta*) in Norway, showed that k and M did not have a positive relation, nor did α and M have a negative one. k/M averaged 0.52 with a large standard deviation (0.55); αM averaged 3.1 but ranged from below 1 to near 5 (with one population at near 9!). While it is

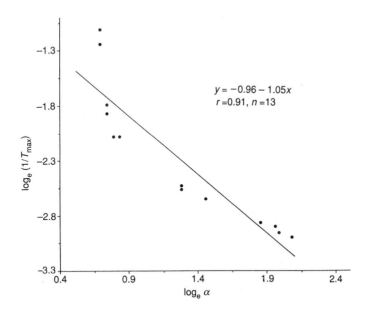

Fig. 4.6 The maximum life-span T_{max} is proportional to the age at maturity α for walleye (*Stizostedion vitreum*) in North America; we have $\alpha/T_{max} = 0.38$. If $M = 6/T_{max}$, we obtain $\alpha M = 2.30$. (Standard errors: slope, 0.15; intercept, 0.20.). Data compiled by Beverton (1987).

Table 4.1. Estimates of parameters of growth, maturity and longevity of walleye at 13 locations, with ambient temperatures.

No.	Location	Temperature* (GDD × 10⁻³)	T_{max} (years)	α (years)	k (years⁻¹)	l_∞ (cm)	l_α (cm)	kT_{max}
1	Big Trout Lake, Ont.	1.04	20	8.0	0.05	85	44	1.0
2	N. Caribou Lake, Ont.	1.16	19	7.3	0.09	72	39	1.7
3	Lac La Ronge, Sask.	1.18	18	7.1	0.06	108	46	1.1
4	Deer Lake, Ont.	1.28	17.5	6.4	0.09	88	38	1.6
5	Escanaba Lake, WI	1.90	14	4.3	0.21	62	41	2.9
6	Lake Winnebago, WI	2.19	13	3.6	0.21	64	39	2.7
7	Pike Lake, WI	2.27	12.5	3.6	0.16	80	41	2.0
8	Current River, MI	3.60	8	2.3	0.22	70	39	1.8
9	Lake Meredith, TX	3.69	8	2.2	0.40	60	46	3.2
10	Center Hill Res, TN	4.08	6.5	2.1	0.32	75	46	2.1
11	El Capitan Res, CA	4.44	6.0	2.1	0.69	62	50	4.1
12	Belton Res, TX	5.30	3.5	2.0	1.19	54	50	3.6
13	Canyon Res, TX	5.68	3.0	2.0	1.10	47	42	3.3

*Sum of the product (days) (degrees above 5°C), where 5°C is the minimum temperature in the growing season which allows positive growth. Analysis from Beverton (1987) based on data from Colby et al. (1979) and Colby and Nepszy (1981).

unclear what to make of this lack of pattern in the data, the αM values are still in the fish range (Fig. 1.10).

4.3   Aquatic invertebrates

Only two invertebrate groups have been studied from the perspective of the Beverton–Holt invariants. The M/k ratio in sea urchins (Echinodermata, class Echinoidea) has been investigated by Ebert (1975), and Charnov (1979c) has compiled data for shrimp in the family Pandalidae. These relations are reviewed in this section.

Sea urchins

Sea urchins are benthic echinoderms, found on hard substrates. Ebert (1975) estimated M and k, from either his own studies or the literature, for 16 populations of 14 species ranging in habitat from tidal to subtidal and in latitude from north temperate to the tropics. Figure 4.7 shows strong proportionality between the two variables with $M/k = 1$, much lower than fish (or shrimp). There is no difference between the tropical

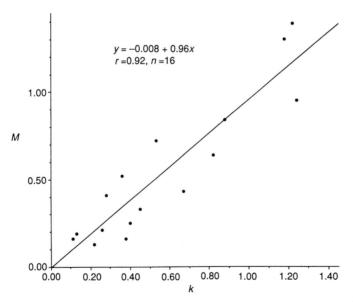

$$y = -0.008 + 0.96x$$
$$r = 0.92, n = 16$$

Fig. 4.7 M versus k in a world-wide survey of sea urchins (16 populations of 14 species). (Standard errors: intercept, 0.075; slope, 0.11). Data from Table 1 of Ebert (1975).

and temperate populations. Unfortunately, there appears to be no comparable data compilation for ℓ_α/ℓ_∞ or αM.

Pandalid shrimp

Shrimp in the family Pandalidae are of major commercial importance in temperate to subarctic waters. They have been the subject of several major symposia and numerous life history monographs and reviews (Butler 1964; Charnov 1979c, 1982; Shumway *et al*. 1985). Most commercial species are protandrous hermaphrodites. They tend to be found at higher temperate and boreal latitudes, associated with mud (or mud plus sand) bottoms at depths ranging from tens to hundreds of metres. Several species show inshore–offshore migrations as well as vertical movement (upward at night) in the water column. Breeding takes place in the autumn. After copulation, females carry the eggs until they hatch in the spring. The larvae are planktonic, settling to the bottom in middle to late summer. Individuals breed for the first time 1.25, 2.25, or 3.25 years after settlement. After a variable time as a male, individuals reverse sex and reproduce as females for the rest of their lives. Some populations have pure females who are not sex-reversed males.

Pandalid shrimp growth is well described by the Bertalanffy equation. Estimates of α, k, and M are available for 27 populations, covering five species, throughout the northern hemisphere (data compiled by Charnov (1979c, 1989a)). Table 4.2 shows the average value of k for each of the three values of α. The product αk is constant (c. 0.82) over the three α values, resulting in the invariance of ℓ_α/ℓ_∞ ($= 0.56$). Figure 4.8 plots $1/T_{max}$ versus k for the shrimp data. The fitted relation is proportional with $kT_{max} = 2.37$. The Beverton (1963) relation $M = g/T_{max}$ was used to estimate M. It was shown earlier that $g \approx 6$ was necessary for shrimp to fit a key prediction from sex allocation theory (Charnov 1979c, 1982). Since $M = 6/T_{max}$, we have $M/k = 2.56$, rather higher

Table 4.2. Relative size at the onset of maturity in pandalid shrimp.

Age at maturity α^* (years)	Average growth coefficient $\overline{k}$	$\alpha\overline{k}$	ℓ_α/ℓ_∞	No. of populations
1.25	0.65	0.81	0.56	14
2.25	0.37	0.83	0.56	11
3.25	0.24	0.78	0.54	2

*Approximate value, measured from larval settlement in summer until autumn breeding under the assumption that the time between summer and autumn is 0.25 year.
Data compiled by Charnov (1979c).

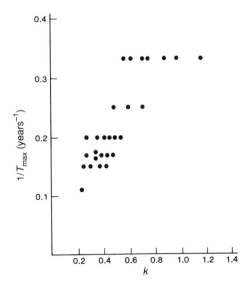

Fig. 4.8 $1/T_{max}$ versus k for 27 populations of five species of pandalid shrimp. The relationship is a simple proportion (by log–log regression) with equation $kT_{max} = 2.37$ ($r = 0.84$, confidence intervals, 2.1–2.7; fit through the origin). The average kT_{max} over the 27 populations is 2.26. Data compiled by Charnov (1979c).

than for fish; however, $\alpha M = (\alpha k)(M/k) = 0.82 \times 2.56 = 2.1$, which is very near the fish values. The α values in Table 4.2 are measured with age zero at larval settlement, and thus may be rather low; upward adjustment of 10–20 per cent have little effect on the general results claimed here.

4.4 Reptiles

Squamate reptiles (snakes and lizards) constitute another group in which growth continues after maturity. Shine and Charnov (1992) compiled data on annual adult survival and ages at female maturation (age zero is hatching) for 16 species of snake (12 Colubrids, four Viperids) from the review by Parker and Plummer (1987, Table 5) and for 20 species of lizards (14 iguanids, three teiids, two lacertids, one xantusiid) from the reviews by Andrews (1982, Appendix 1), Dunham *et al.* (1988, Appendix), and Shine and Schwarzkopf (1992, Table 1, and references cited therein). There is a strong phylogenetic and geographical bias in the species studies (mostly North American iguanid lizards and colubrid snakes). The compilation included data on separate populations of two wide-ranging lizard species (*Sceloporus graciosus* and *Sceloporus undulatus*) and one snake (*Crotalus viridis*).

Reptiles differ from most fish and pandalid shrimp in one important respect: they start life (size at hatching) at a much larger relative size, making the two-parameter Bertalanffy equation an inadequate model for growth. Figure 4.9 shows the three-parameter form where

$$\ell_x = \ell_\infty \{1 - \exp[-k(x - x_0)]\} \ .$$

At hatching, i.e. time zero ($x = 0$), the individual is of relative size $\ell_0/\ell_\infty = 1 - \exp(kx_0)$. The growth curve is only fitted with times greater than zero and hence the x_0 parameter is merely a device that allows the size at hatching to be made substantially greater than zero; x_0 gives a zero time which would result if growth from size zero also followed the curve fitted for post-hatching growth. Actually, growth prior to hatching is much faster than this.

Shine and Charnov (1992) also compiled data on length ℓ_α at sexual maturation and the maximum length. For large samples this maximum length is known to be about 5 per cent less than ℓ_∞ (Taylor 1962; Pauly 1981). Shine (unpublished) has compiled estimates of the relative size ℓ_0/ℓ_α at hatching. Shine and Charnov (1992) used the two-parameter Bertalanffy curve to estimate the k values, resulting in relatively large overestimates. Here the ℓ_α/ℓ_∞ and ℓ_0/ℓ_α data are used to provide

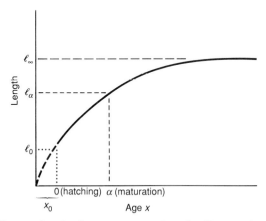

Fig. 4.9 Reptiles require the three-parameter Bertalanffy growth equation since at age zero (hatching) they are usually 40–50 per cent of their maturation length. The equation is

$$\ell_x = \ell_\infty \{1 - \exp[-k(x - x_0)]\} \ .$$

better estimates of the k values (method described in the caption to Fig. 4.11).

ℓ_0/ℓ_α is relatively invariant within lizards and snakes and about 20 per cent different between the two groups (0.48 for lizards versus 0.40 for snakes). The standard error is 0.01 for both groups (17 lizards,

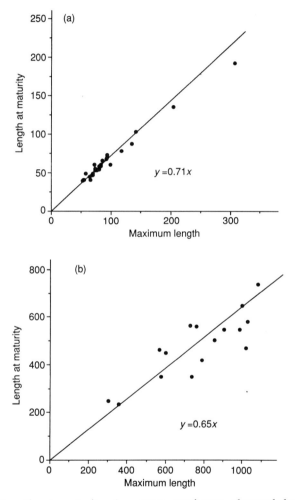

Fig. 4.10 Length at maturity ℓ_α versus maximum observed length for (a) 28 lizard populations (20 species) and (b) 17 snake populations (16 species). There is a small negative allometry with larger individuals relatively smaller at maturation. Lines through the origin yield a ratio of 0.65 (standard error, 0.025) for snakes and 0.71 (standard error, 0.011) for lizards. Since ℓ_∞ is about 5 per cent larger than the maximum length, ℓ_α/ℓ_∞ is about 0.62 for snakes and 0.68 for lizards. Data compiled by Shine and Charnov (1992).

28 snakes). Figure 4.10 plots ℓ_α versus maximum length for both groups; there is a slight negative allometry (first described by Andrews (1982)) so that larger species tend to mature at a relatively smaller size. This effect is not very large, however, and the figure shows the fitted proportional relations. Lizards have the higher slope (0.71 versus 0.65) but these differences are not statistically significant. ℓ_α/ℓ_∞ is about 0.95 times the respective slope. Figure 4.11 shows $\log_e M$ versus $\log_e k$; the relation has a slope near unity, making M/k a constant equal to 1.5 which is not significantly different from the value for fish ($M/k = 1.65$, Fig. 4.5).

Figure 4.12 shows $\log_e M$ versus $\log_e \alpha$ for reptiles. Again, the slope is not different from -1, making αM a constant equal to 1.32. As

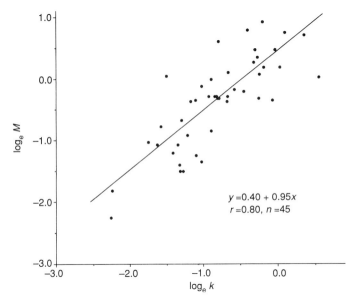

$$y = 0.40 + 0.95x$$
$$r = 0.80, \ n = 45$$

Fig. 4.11 $\log_e M$ versus $\log_e k$ for 45 snake and lizard populations. The M/k ratio is a constant equal to $e^{0.40} = 1.5$, not very different from the fish of Fig. 4.5. (Standard errors: slope, 0.11; intercept, 0.11.) The average M/k ratio is 1.73 (standard error, 0.12). k is estimated as follows with reference to Fig. 4.9. Shine and Charnov (1992) and Shine (unpublished) give data on ℓ_α, $\ell_{max} (= 0.95 \ell_\infty)$, ℓ_0 and α for the various species/populations. For the three-parameter Bertalanffy equation, we have

$$\frac{\ell_\alpha}{\ell_\infty} = 1 - \exp[-k(\alpha - x_0)] \quad \text{and} \quad \frac{\ell_0}{\ell_\infty} = 1 - \exp(kx_0) .$$

With ℓ_α, ℓ_∞, ℓ_0 and α known for each population, we have two equations and two unknowns (k, x_0) for each population and thus we can calculate rough estimates of k and x_0. Of course, more precise estimates of k and x_0 would follow from more complete size at age data.

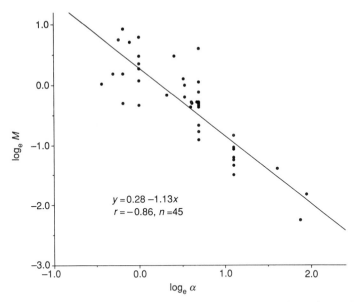

$y = 0.28 - 1.13x$
$r = -0.86, n = 45$

Fig. 4.12 M is inversely proportional to the age at maturity α in snakes and lizards; $\alpha M = 1.32$ (standard errors: slope, 0.10; intercept, 0.08). Data compiled by Shine and Charnov (1992).

noted earlier, reptiles differ from most fish in having large offspring at hatching ($\ell_0/\ell_\alpha \approx 0.4-0.5$). While the growth curve is really only defined post-hatching, it is interesting to ask what αM would look like if α were measured from size zero as given by the backward extrapolation of the growth curve; here age at maturity is redefined to be $\alpha^* = \alpha - x_0$. The x_0 values are estimated together with the k values (see caption to Fig. 4.11). If we now fit $\log_e M$ versus $\log_e \alpha^*$, the relation is $y = 0.76 - 1.17x$. The slope is not significantly different from -1, making αM an invariant equal to 2.1, virtually the same as for fish and shrimp. It is as if the only major difference between the groups is the initial condition—the size at hatching. Reptiles spend only a few months in the egg stage and so they reach the size at hatching in a much shorter time than would follow from extrapolation of the Bertalanffy curve back to size zero. Indeed, the reptile data show $-x_0/\alpha \approx 0.67$, whereas the egg stage may well account for less than 10 per cent the time from hatching to maturation. Perhaps the difference between these two numbers points to a major advantage of development within a reptile egg.

4.5 A life history theory for the Beverton–Holt invariants

Beverton (1963) went beyond the basic description of the patterns and proposed that they must reflect the action of natural selection. In his words:

... An adaptive interpretation of the adjustment of growth and longevity might be explicable in simple Darwinian terms by supposing that it is an 'advantage' to the individual to complete as much as possible of its potential growth within its likely maximum life-span, this advantage being that which best enables its reproductive potential to be realized and so maximizes the contribution of its progeny to future generations. But in this case the age at first maturity must also be adjusted to life-span; that this happens is shown in [his] Figure 7 by the constancy of the ratio of length at first maturity to asymptotic length $\ell_\alpha / \ell_\infty$.

In this section we build on Beverton's pioneering suggestion and ask evolutionary theory to output the Beverton–Holt patterns (invariance in M/k, ℓ_α/ℓ_∞, αM). Three basic assumptions are made (Charnov 1986, 1990; Charnov and Berrigan 1991a,b; Chapter 1 of this book).

1. Selection is stabilizing in a stationary population so that the net reproductive rate R_0 is an appropriate fitness measure. R_0 is maximized with the population dynamic constraint that $R_0 = 1$; this is implicitly enforced through density-dependent mortality on young immatures. The individual's growth function and the adult mortality rate are independent of population size. Consider a newborn female and define $L(x)$ as the probability she is alive at age x, and $b(x)$ as her birth rate, in daughters, at age x. Her life-time production of daughters is $R_0 = \int_\alpha^\infty L(x)b(x)\,\mathrm{d}x$.

We can rewrite R_0 as

$$R_0 = L(\alpha)\left[\frac{\displaystyle\int_\alpha^\infty L(x)b(x)\,\mathrm{d}x}{L(\alpha)}\right].$$

The term in square brackets is the average number of daughters born over a female's adult life-span, the 'Fisherian reproductive value' (Fisher 1930) of a just mature female of age α, and will therefore be labelled $V(\alpha)$. Thus we have $R_0 = L(\alpha)V(\alpha)$. To allow for the fact that mortality usually decreases over the immature period, write $L(\alpha)$ as:

$$L(\alpha) = \exp\left[-\int_0^\alpha Z(x)\,dx\right]$$

where Z for x near zero is the sole source of density dependence in the population (which acts to make $R_0 = 1$). Z, near and after α, is assumed to be constant (i.e. type II mortality).

2. $V(\alpha)$ is a function of body size at maturity, so that $V(\alpha) = D_1 \ell_\alpha^p$ where ℓ_α is the length at age α. This assumption aggregates reproduction over the adult life-time by suggesting that $V(\alpha)$ can often be at least approximated as a power function of adult body size at maturity. This assumption is an attempt to avoid explicit modelling of life-time reproduction under indeterminate growth.

3. Finally, it is assumed that a trade-off exists for the growth curve as illustrated in Fig. 4.13; an individual is allowed fast early growth with a small maximum size, or slow early growth with a large maximum size. Growth is assumed to follow a Bertalanffy curve: $\ell_x = \ell_\infty(1 - e^{-kx})$. It should be noted that for x near zero this growth equation is approximately $\ell_x = \ell_\infty[1 - (1 - kx)]$ or $\ell_x = \ell_\infty kx$. Thus the growth rate $d\ell_x/dx$ for small x is $k\ell_\infty$. Now, suppose that ℓ_∞ and k are negatively related by the (trade-off) function $\ell_\infty = Dk^{-h}$ (or $k = (D/\ell_\infty)^{1/h}$). This yields a growth rate for small x of $k\ell_\infty = D^{1/h}\ell_\infty^{-1/h}\ell_\infty = D^{1/h}\ell_\infty^{(1-1/h)}$. This early growth rate $k\ell_\infty$ is negatively

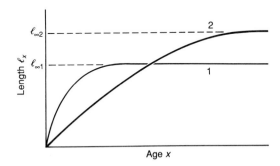

Fig. 4.13 A growth rate trade-off exists if fast early growth *must* be followed by a small asymptotic size: curve 1 versus curve 2. For the Bertalanffy curve $\dfrac{d\ell_x}{dx} \approx k\ell_\infty$ for ℓ_x near zero, and so a trade-off follows if ℓ_∞ and k are negatively related by $\ell_\infty \propto k^{-h}$ with $0 < h < 1$.

related to ℓ_∞ if $1 - 1/h < 0$ or $h < 1$; high early growth ($k\ell_\infty$ high) must be followed by small asymptotic size ℓ_∞ if $h < 1$. This is what is assumed for members of a population $\ell_\infty = Dk^{-h}$ with $h < 1$.

With these assumptions R_0 can now be written as $R_0 = L(\alpha)V(\alpha)$, where

$$L(\alpha) = \exp\left[-\int_0^\alpha Z(x)\,\mathrm{d}x \right] \quad \text{and} \quad V(\alpha) = D_1 \ell_\alpha^p .$$

ℓ_α is given by $\ell_\alpha = \ell_\infty (1 - e^{-k\alpha})$; recall the growth trade-off ($\ell_\infty = Dk^{-h}$) so that $\ell_\alpha = Dk^{-h}(1 - e^{-k\alpha})$ or $V(\alpha) = D_1 \ell_\alpha^p = D_1 D^p k^{-hp}(1 - e^{-k\alpha})^p$. Finally, we have

$$R_0 = \exp\left[-\int_0^\alpha Z(\alpha)\,\mathrm{d}x \right] D_1 D^p k^{-hp}(1 - e^{-k\alpha})^p$$

or

$$\log_e R_0 = \log_e (D_1 D^p) + p\log_e (1 - e^{-k\alpha}) - ph\log_e k - \int_0^\alpha Z(x)\,\mathrm{d}x .$$

Natural selection acts on this life history through choice of α and k. Evolutionary equilibrium occurs when $\partial\log_e R_0/\partial\alpha = 0$ and $\partial\log_e R_0/\partial k = 0$ simultaneously. In what follows, M refers to the adult instantaneous mortality rate $Z(\alpha)$, which is assumed to be constant over the adult life-span (it stops decreasing at age α) and R refers to the relative length at the age at maturity ($R = \ell_\alpha/\ell_\infty = 1 - e^{-k\alpha}$). Setting these two derivatives equal to zero produces the following results:

$$\left[\frac{\partial\log_e R_0}{\partial k} = 0 \right] \quad h = \frac{R-1}{R}\log_e(1 - R) \tag{4.2}$$

$$\left[\frac{\partial\log_e R_0}{\partial\alpha} = 0 \right] \quad p\frac{k}{M} = \frac{R}{1 - R} . \tag{4.3}$$

These two equations, as functions of the two shape coefficients p and h, thus fix the value of two *dimensionless numbers* k/M and

$\ell_{\alpha}/\ell_{\infty}$ $(=R)$. Four implications of these equations are noteworthy. First, the height coefficients D and D_1 do not appear in the solutions; *the predicted values of k/M and $\ell_{\alpha}/\ell_{\infty}$ are invariant to alterations (transitions) in height.* As long as shape (h and p) is maintained, as illustrated in Fig. 4.14, the same k/M and $\ell_{\alpha}/\ell_{\infty}$ are predicted. Second, from eqn. (4.2), only one $\ell_{\alpha}/\ell_{\infty}(=R)$ is allowed for a given shape h. Third, eqn. (4.3) shows that for given R and p values, only one k/M ratio is allowed; thus all species with the same h and p values are predicted to have the same k/M ratio. Fourth, rewrite eqn. (4.3) as

$$ p = \frac{R}{1 - R} \frac{M}{k} $$

and multiply it by eqn. (4.2) to yield

$$ ph = \frac{-M}{k} \log_e (1 - R) \quad . $$

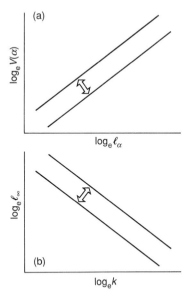

Fig. 4.14 Under the life history theory developed here, the dimensionless numbers k/M, $\ell_{\alpha}/\ell_{\infty}$, and αM are invariant to height changes in the relations between $\log_e V(\alpha)$ versus $\log_e \ell_{\alpha}$ and the trade-off between $\log_e \ell_{\infty}$ and $\log_e k$. The numbers depend only upon the *slopes* (the *shapes*) of these relations and not on their height.

Now, $R = \ell_\alpha/\ell_\infty = 1 - e^{-k\alpha}$ so that $-\log_e (1 - R) = k\alpha$. Thus

$$ph = \alpha M ,$$

i.e. the adult instantaneous mortality rate M is inversely proportional to the age at maturity α with the proportionality constant equal to ph.

Figure 4.15 shows a graphical solution to eqns. (4.2) and (4.3), and illustrates the observed k/M region of 0.5–0.65. In general, $p \approx 3$–5 is required to produce this output. This p hypothesis, not pursued further here, is that life-time egg production will scale as a power function of length at maturity ℓ_α with an exponent of c. 3–5 (or will scale with weight at maturity to the power 1.0–1.67).

It should now be clear what meaning can be attached to the negative relation often shown between $\log_e \ell_\infty$ and $\log_e k$. Species or populations with the same h and similar D values should fall on the same curve; this illustrates the growth rate trade-off. If logarithmic regressions (functional regressions, (Ricker 1973)) of ℓ_∞ on k can be used to estimate

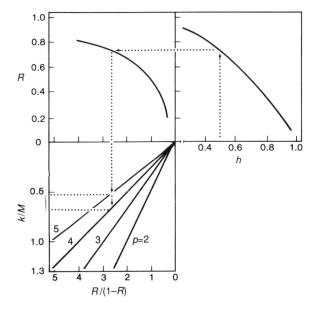

Fig. 4.15 A graphical method of estimating the equilibrium k/M and $\ell_\alpha/\ell_\infty (=R)$ from p and h (or vice versa) based on eqns. (4.2) and (4.3) in the text. The right-hand panel is the solution of eqn. (4.2) (R as a function of h). The upper left-hand panel transforms R into $R/(1-R)$. The lower left-hand panel is the solution to eqn. (4.3); it transforms $R/(1-R)$ into k/M for the appropriate p value. The dotted lines show the k/M range of 0.5 to 0.65 for $h = 0.5$.

h, then the Pauly analysis mentioned earlier in this chapter gives about the right magnitude for h (c. 0.5) since most fish have $\ell_\alpha / \ell_\infty$ in the range 0.4–0.8 (Beverton and Holt 1959), leading through eqn. (4.2) to h in the range 0.4–0.8 (Fig. 4.15). Plots of $\log_e \ell_\infty$ versus $\log_e k$ for six species (two cod, two engraulids, and two clupeids; data compiled by Pauly (1978)) are shown in Fig. 4.16 to illustrate the relations often seen with large sample sizes ($n > 12$). This regression technique should give a reasonable estimate of h if (1) the data come from populations with similar $\ell_\alpha / \ell_\infty$ ($= R$) values and (2) the height parameters of the particular trade-offs (the D of $\ell_\infty = Dk^{-h}$) are the same and/or at least uncorrelated with ℓ_∞ or k. Of course, a negative relation will be seen only if ℓ_∞ and k span a large range relative to the range of the respective D values; otherwise we may even see a positive correlation between k and ℓ_∞ (see van Noordwijk and de Jong (1986) for further discussion of this point for trade-offs in general or Fig. 4.8 of Stearns (1992)).

Recall from eqn. (4.2) (or Fig. 4.15) that h is predicted to be inversely related to R. A test of this h versus R hypothesis would be to estimate the slope of the functional regression of $\log_e \ell_\infty$ on $\log_e k$ within groups of species (or populations) with similar R values, and then see if the relation between R and h looks like eqn. (4.2). One comparison of this type is possible using the growth data compilation of Pauly (1978). He compiled or estimated growth curves for 1501 populations, stocks, or year classes of 515 species of fish in 104 families. Beverton and Holt (1959) and Beverton (1963) showed that R averaged approximately 0.55 for the cod family (Gadidae) and approximately 0.75 for the herring, sardines, and anchovies (the Clupeomorpha, families Clupeidae and Engraulidae); these values for R are near the outer bounds for fish. Equation (4.2) predicts $h \approx 0.45$ for fish in the Clupeomorpha and $h \approx 0.65$ for cod. In the Pauly (1978) compilation 11 species (three cods, five clupeids, and three engraulids) have 13 or more observations per species (stocks and/or year classes); Table 4.3 shows the results of functional regressions of $\log_e \ell_\infty$ on $\log_e k$ within each of these species. The estimates of h range from 0.27 to 0.75; Fig. 4.17 shows h broken down by cod versus Clupeomorpha. The cod are generally much higher than the Clupeomorpha (the two highest h values are for cod) and the averages $\bar{h}$ are not far from the theoretical predictions; cod, 0.57 (observed) versus 0.65 (predicted); Clupeomorpha, 0.40 (observed) versus 0.45 (predicted). While this comparison involves few species (and while the within-family/ between-species differences are not explained), the averages do indeed provide support for the theoretical approach suggested here. Beverton (1992) has recently suggested that R averages closer to 0.60 for cod and 0.80 for clupeids; these R values produce theoretical h values (0.61, and 0.40) even closer to the observed averages of 0.57 and 0.40.

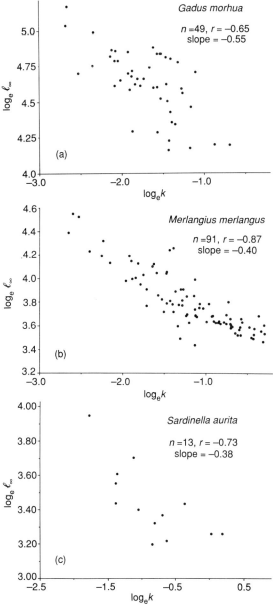

Fig. 4.16 $\text{Log}_e \ell_\infty$ versus $\log_e k$ within each of six fish species: (a), (b) cod; (c), (d) clupeids; (e), (f) engraulids. Each data point represents a separate population, stock, or year class. Slope refers to a functional regression (statistics in Table 4.3). The functional regression assumes equal error variance on each axis; if more precise information is known about the error variances, other estimation schemes become more reasonable (Ricker 1973; Pagel and Harvey 1988).

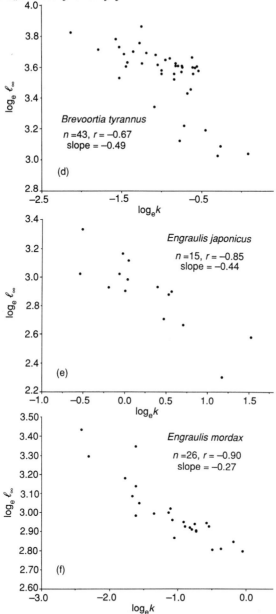

Fig. 4.16 (*continued*) The use of functional regression is motivated by the fact that between-species (or between-population) variation in D (or $\log_e D$) contributes to error almost symmetrically on both axes. It should also be noted that the growth data used to estimate ℓ_∞ and k (within each population) must span the entire age range; otherwise ℓ_∞ and k will have a large negative covariance simply due to the estimation of each from the same data. Data compiled by Pauly (1978).

Table 4.3. Estimates of h from within-species regressions of $\log_e \ell_\infty$ on $\log_e k$.

Family/species	h^*	Standard error of h	Correlation coefficient (sample size)[†]	Probability level (two-tailed test)
Gadidae (Cod)				
Gadus morhua	0.55	0.06	$r = -0.65$ ($n = 49$)	< 0.0001
Melanogrammus aeglefinus	0.75	0.14	$r = -0.78$ ($n = 13$)	< 0.002
Merlangius merlangus	0.40	0.02	$r = -0.87$ ($n = 91$)	< 0.0001
Clupeidae (herring, sardines)				
Clupea harengus	0.50	0.08	$r = -0.39$ ($n = 39$)	< 0.014
Sardina pilchardus	0.42	0.08	$r = -0.66$ ($n = 18$)	< 0.003
Sardinops caerula	0.41	0.08	$r = -0.54$ ($n = 19$)	< 0.02
Sardinella aurita	0.38	0.08	$r = -0.73$ ($n = 13$)	< 0.005
Brevoortia tyrannus	0.49	0.06	$r = -0.67$ ($n = 43$)	< 0.0001
Engraulidae (anchovies)				
Engraulis japonicus	0.44	0.065	$r = -0.85$ ($n = 15$)	< 0.0001
Engraulis encrasicholus	0.31	0.06	$r = -0.66$ ($n = 17$)	< 0.004
Engraulis mordax	0.27	0.025	$r = -0.90$ ($n = 26$)	< 0.0001

$*$ h is minus the slope of the functional regression (major axis regression) of $\log_e \ell_\infty$ versus $\log_e k$ within a species or between populations and/or year classes (see second footnote). The within-species data were used to estimate the h values because plots at higher taxonomic levels (e.g. within a family or pooling all species) lead to higher values of h. This effect probably occurs because larger species have higher D values (but it could also be a statistical artefact (Pagel and Harvey 1988)).

[†] A datum is a separate population, stock, or year class.

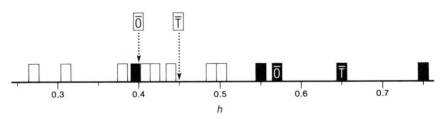

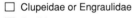

Fig. 4.17 A graphical representation of the *h* values of Table 4.3. In general, the cod (■) are higher than the Clupeomorpha (□); the averages ($\overline{O}$, observed; $\overline{T}$, theoretical) are also shown.

4.6 Summary

The Beverton–Holt invariants (the dimensionless numbers M/k, ℓ_α/ℓ_∞, and h) have gone unnoticed by the ecological community for nearly 30 years. If the data patterns continue to hold for other organisms with indeterminate growth, they strongly suggest that evolution is governed by some *very general features of life history trade-offs*. Our task is to find them, and principles of symmetry remain our guide, just as Fisherian invariance is our guide for general features of sex allocation. Lack of invariance within the walleye and the brown trout should also be examined.

The theory developed in this chapter aggregates the life history using the growth rate trade-off ($\ell_\infty \propto k^{-h}$) while assuming that we can map size at maturity into adult reproductive success $V(\alpha) \propto \ell_\alpha^p$. Some have questioned the Bertalanffy equation as an adequate descriptor of fish (or reptile) growth (Roff 1980; Andrews 1982). Its use here is motivated more by the fact that much of the published literature estimates its parameters than by any feeling for its biological basis; it frequently fits the age size data quite well. We could use dynamic programming techniques and aim to output the allocation of energy to growth versus reproduction as a function of age. This approach would derive the growth and maturation relations as automatic consequences of these more basic allocation decisions, including mortality rates and trade-offs, in the face of specified production rates (Sibly *et al.* 1985; Kozlowski 1992; Vance 1992). Such models usually present formidable mathematical problems, particularly for the case of indeterminate growth. Therefore, a more descriptive (aggregate) approach is used in this chapter.

We shall return to the case of indeterminate growth again in Chapter 8, to develop a hypothesis about fish growth which produces *approximately* the Bertalanffy equation (Roff 1983, 1984). This new approach uses a body size production model, introduced in Chapter 5, and gives rise to a new dimensionless number relating reproductive effort to the age at maturity, a number which is invariant for constant $\ell_\alpha / \ell_\infty$ values. The reason for delaying this production-based approach until the end of the book is that it is most easily developed and understood for determinate growers ($\ell_\alpha / \ell_\infty \approx 1$), as will be done for mammals in the next chapter.

5

Determinate growth, mostly about mammals

We suggest, then, that the scaling of life history variables with body weight may derive from more fundamental relationships of weight with mortality schedules. But why should there be any consistent relationship between weight and rates of mortality? We do not yet know the answer to this question ... This approach, of attempting to explain allometric relationships as a function of demographic reality, is, we think, preferable to the largely descriptive role that allometry has played over the years. (Harvey and Pagel 1991, p. 198). (An answer to this question is provided by eqn. (5.7).)

5.1 Introduction

Birds, mammals, and insects show determinate growth where adult weight is more or less unchanging. Birds commonly reach their adult size well before maturation, while insects and mammals mature upon reaching near adult size. This chapter mostly deals with the allometry (body size scaling) of life histories for female mammals and the resulting invariants which follow from shared scaling exponents. Recall that if $Y_1 \propto W^P$ and $Y_2 \propto W^{-P}$, the product $Y_1 Y_2$ will be independent of body size W: $Y_1 Y_2 \propto W^P W^{-P} \propto W^0$. At the end of the chapter the evolution of sexual dimorphism (asymmetry) in adult body size is briefly discussed.

Two major approaches have dominated recent thinking about large-scale variation in mammalian life histories. The first sees variation as simply reflecting allometric or scaling consequences of adult body size (Millar 1977; Western 1979; Western and Ssemakula 1982; Peters 1983; Calder 1984; Reiss 1989), while the second sees natural selection as moulding life history fairly independent of body size (Harvey and Zammuto 1985; Harvey *et al.* 1989; Read and Harvey 1989; Promislow and Harvey 1990; Harvey and Pagel 1991). Both approaches recognize that 'to grow big takes time' so that larger adults must have longer immature periods; they differ in that the first assumes that death rates and birth rates are also mainly determined by adult size, with a rather mysterious causal connection (at least for death rates (Linstedt and Calder 1981; Linstedt and Swain 1988)), while the second sees these rates as free to evolve within broad limits independent of adult size. A major finding supporting the second position is that life history variables like birth rates, death rates, and age at maturity are highly correlated with each other even when adult body size is held constant (see the papers by Harvey and co-workers cited above).

This chapter unifies the two approaches (Charnov 1991). It assigns a central role to body size. This role is a 0.75 scaling for net production which results in two things: first, adult body size is simply the result of growth, accumulated production prior to age α; second, offspring production is simply net production diverted from personal growth. The theory also invokes natural selection on the age at maturity to link adult demography to adult body size through the effects of size on individual productivity (or growth potential). Finally, the approach makes use of the demographic identity appropriate for a non-growing population; the net reproductive rate R_0 must equal unity in such a population so that not all variables in R_0 can vary independently (Calder 1984; Charnov 1986; Sutherland *et al.* 1986). The causal chain is illustrated in Fig. 5.1 (Harvey and Nee 1991).

5.2 Empirical patterns for female mammals

Several life-history variables scale as power functions of adult body size. In particular, age at maturity α, life expectancy at birth, life expectancy at maturity, and annual fecundity b all scale with exponents p, where $Y \propto \text{weight}^p$, near ± 0.25 (fecundity is negative) (Stahl 1962; Linstedt and Calder 1981; Millar and Zammuto 1983; Peters 1983; Calder 1984; Wooten 1987; Harvey *et al.* 1989; Read and Harvey 1989; Reiss 1989; Promislow and Harvey 1990) (see Fig. 1.13). The extensive analyses are at a variety of taxonomic levels and degrees of precision. The exponents invariably fall in the range 0.2–0.35. For example, the average instantaneous mortality rates $\overline{Z}$ and M for the juvenile and adult periods of life respectively were estimated using the life-table data set compiled by Millar and Zammuto (1983). Both scaled with exponents near -0.25 (adult: -0.21, $r = 0.83$, $n = 26$; juvenile: -0.27, r $= 0.89$, $n = 26$). This scaling is expected since life expectancy is the inverse of the mortality rates (and it scales with a $+0.25$ power). Wooten (1987) compiled a sample of 547 mammalian species; α scaled with an exponent of 0.25 ($r = 0.75$). Of course, shared scaling exponents means invariance for the dimensionless relations αM, αb, and $\alpha \overline{Z}$, and one aim is to predict their numerical values from theory.

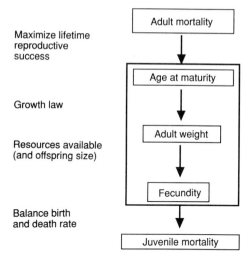

Maximize lifetime
reproductive
success

Growth law

Resources available
(and offspring size)

Balance birth
and death rate

Fig. 5.1 A causal chain for mammalian life history evolution. Adult mortality is determined by the environment. Mammals mature at the age which maximizes life-time reproductive success. A growth law determines adult size as a function of age at maturity. At maturity, the energy which would have gone into growth is channelled into the production of offspring. In the long term, the birth rate must equal the death rate in natural populations. This equality is ensured by juvenile mortality's being density dependent: at low population densities fecundity exceeds mortality, but at high density mortality exceeds fecundity. This assumption about density-dependent juvenile mortality is supported by data reviewed by Fowler (1981, 1987) and Sinclair (1989). From Charnov (1991) and Harvey and Nee (1991).

Harvey and colleagues have also shown that these life history variables (age at maturity, adult and juvenile mortality rates, and annual fecundity) are correlated with each other even when adult body size is held constant. Table 5.1 shows the results of the following analyses, which are illustrated in Fig. 5.2. The authors fit lines of the form $\log_e$ (variable) versus $\log_e$ (adult weight). Since some taxa fell above or below the line, the deviation (from the line) of the particular $\log_e$ (variable) can be calculated. These residuals are highly correlated with each other (Table 5.1). For example, taxa with relatively late ages at maturity also have relatively low adult mortalities.

5.3 Theory: the basic 0.25 scaling

Stable demography

Consider a life history where growth ceases at adulthood, and which is characterized by constant fecundity and constant adult mortality. Let b

Table 5.1. Signs of correlations between life history variables for mammals, with adult body weight held constant, as illus trated in Fig. 5.2.

Variable pair [*]	Observed sign of correlation [*]
Adult mortality rate M [†] Age at maturity α	−
Juvenile mortality rate $\overline{Z}$ Age at maturity	−
Juvenile mortality rate [†] Adult mortality rate	+
Annual fecundity b Age at maturity	−
Annual fecundity Adult mortality	+
Annual fecundity Juvenile mortality	+

[*] Each pair is expressed in $\log_e$ ( ) form; the observed correlation is between respective deviations of the variables from their means, derived as the deviations (or residuals) from the $\log_e$ ( ) versus $\log_e$ (adult body weight) regression. Data compilation and analysis from Harvey and Zammuto (1985); Harvey *et al.* (1989); Read and Harvey (1989); Promislow and Harvey (1990); the annual fecundity b is estimated as an energetic maximum, not as a field observation.

[†] The exact measures of adult and juvenile mortality rates used by Promislow and Harvey (1990) differ from the instantaneous rates developed in the theory in this book. To see whether these somewhat different measures affected the signs of the correlations, the $\log_e$ deviation analysis was repeated using the appropriate instantaneous rates calculated for the life-table data compilation of Millar and Zammuto (1983). In terms of the signs of the correlations, the more precise analysis used here gave the same answers.

be the birth rate in daughters per unit time, $S(\alpha)$ the survival fraction of daughters to maturity (age α), and M the adult instantaneous mortality rate. R_0 is given by (as derived in Chapter 1, eqn. (1.2) with $V(\alpha) = b/M$)

$$R_0 = \frac{bS(\alpha)}{M} \quad . \tag{5.1}$$

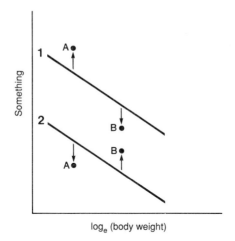

Fig. 5.2 A negative correlation between the residuals of (deviations from) two life history allometries. The fitted allometries for variables 1 and 2 include many separate species, but here we focus on only two, A and B. Species A falls above the line for variable 1 but below the line for variable 2. Species B shows just the reverse behaviour. Positive (negative) deviations from allometry 1 are always matched by negative (positive) deviations from allometry 2. Thus a correlation calculated between the respective deviations will be negative. If positive (negative) deviations from allometry 1 are matched by positive (negative) deviations from allometry 2, the sign of the correlation between respective deviations would be positive.

In a non-growing population $R_0 = 1$ so that any two of the variables set the value of the third. It is often useful to write $S(\alpha) = \exp\left[-\int_0^\alpha Z(x)\,\mathrm{d}x\right]$ where $Z(\alpha)$ is the immature instantaneous mortality rate appropriate for age x. In general, $Z(x)$ decreases with x but it may well reach its lowest value prior to age α. Density dependence is assumed only to affect $Z(x)$ for small x.

Growth versus body size

Several authors (Kozlowski and Wiegert 1986, 1987; Reiss 1989) have noted that the scaling between adult body size and age at maturity (a 0.25 power) follows directly from a growth relation provided that the growth rate (individual productivity) scales with the 0.75 power of body size after independence from the parent. There is evidence that individual productivity, the difference between intake and respiration, does scale in approximately this way within many animal groups (Farlow 1976;

Blueweiss *et al.* 1978; Case 1978; Lavigne 1982; Reiss 1989). Supporting data include both between-species allometries and individual growth functions (Case 1978). If W is body weight, we have

$$\frac{dW}{dT} = AW^{0.75} .$$ (5.2)

The coefficient A differs between taxa, but the exponent, shown here as 0.75, is generally between 0.65 and 0.85. If we take time zero as the point of independence from the parent (approximately the time of weaning) and let W_0 be the corresponding offspring size, eqn. (5.2) can be integrated to give:

$$W(T)^{0.25} = 0.25AT + W_0^{0.25} .$$

If α is the age at maturity (measured from independence) when growth ceases, and if $W_0/W(\alpha)$ is defined as δ, the above equation can be written as

$$W(\alpha)^{0.25} = \frac{0.25A}{1 - \delta^{0.25}} \alpha .$$ (5.3)

Thus the age at maturity α will scale with the 0.25 power of adult body size among species with similar A and δ values. Of course, species may fall above or below the average scaling line depending upon their individual A and δ values, as conditioned by phylogeny (see below) or perhaps ecology (Wooten 1987); this type of prediction will also be possible for all the other allometries derived here. The relative size at independence δ is the point at which an animal moves from a potential growth trajectory determined by its primary care-giver to one determined by its own size. Thus, it is assumed that a mammal grows during two periods: one from birth to independence, where its mother controls its growth rate, and the second from independence to maturity, where its own size determines its growth rate. The size at maturity is assumed to be the final adult size, and this size, in turn, determines the offspring production rate through eqn. (5.2). The age at maturity α is the interval between independence (estimated here as weaning) and first reproduction.

What is the value of A in eqn. (5.2) for mammals (on average)? A has units of $W^{0.25}$ (time)$^{-1}$. With W in kilograms and time in years, A averages approximately unity by three independent estimates, as documented in Table 5.2. This average A was obtained for several rather

Table 5.2. Three independent ways of estimating the value of A eqn. (5.2) for female mammals. *

1(a). Regression of $\log_e \alpha$ on $\log_e W$ (data from Millar and Zammuto (1983) excluding primates). The intercept is 0.06 ($r = 0.85$; sample size, 26; standard error of intercept, 0.10). By eqn. (5.3) with $\delta = 0.33$ (from Fig. 5.4), this intercept should equal $-\log_e A$. Thus $A = 0.94$.

1(b). Regression of $\log_e \alpha$ on $\log_e W$ (data from Wooten (1987) for 547 mammalian species, including those in 1(a) and a very few primates). He does not report the intercept but the slope is 0.25 ($r = 0.75$). From his Fig. 1 the intercept is estimated to be about 4.1 with units of grams and days, or about -0.07 for kilograms and years. Again, by eqn. (5.3) with $\delta = 0.33$, this is $-\log_e A$. Thus $A = 1.07$.

2. Regression of $\log_e M$ on $\log_e W$ (data from Millar and Zammuto (1983)). The intercept is -0.42 ($r = 0.83$; sample size, 26; standard error of intercept, 0.12). By eqn. (5.7) this intercept should equal $\log_e 0.75A$. Thus $A = 0.88$.

3. For a numerically stable population, the ratio of population production to population biomass should equal M. Farlow (1976) shows a -0.27 scaling of this ratio (or M) for 62 populations of non-primate mammals ($r = 0.93$). The intercept of the log–log plot is -0.04: thus by eqn. (5.7) $A = 1.23$. Banse and Mosher (1980) rejected all but seven of Farlow's data points; their scaling relation had a slope of -0.33 ($r = 0.98$) but almost the same intercept.

4. *Primates are very different*. Regression of $\log_e \alpha$ and $\log_e W$ (data from Ross (1992) for 72 primate species; slope is 0.34 ($r = 0.88$; standard error, 0.02)). The intercept is 0.87 (standard error, 0.04), so that by eqn. (5.3) with $\delta = 0.33$ (Fig. 5.4), we have $-\log_e A = 0.87$. Thus $A = 0.42$.

---

* It is important that the allometries for α, M, and the production-to-biomass ratio lead to similar A values, since each relies on different assumptions to relate A to the allometry. Units are kilograms and years unless stated otherwise.

heterogeneous collections of mammals, excluding primates. Primates have very different A values from this average, and are estimated to have $A = 0.42$ (by the methods of Table 5.2). Undoubtedly, other mammal taxa also have A larger/smaller than the average of unity. This estimation relies partly on equations derived later in this chapter.

Offspring production versus adult body size

Although a fair amount is known about offspring growth and energetics during the period of parental care (Brody 1964; Case 1978; Calder 1984;

Reiss 1989) a very simple aggregated model will be used for this process. The clutch size in daughters per unit time is denoted by b. It should be noted that b is the ratio of the brood size to the average time between broods, so that neither appears alone in the theory. Read and Harvey (1989) call b 'annual fecundity'. Suppose that at independence each offspring is of size W_0 and that rearing an offspring to this size requires energy $(C/2)W_0$, where C is in relative units of parental growth. Thus an offspring of size W_0 at weaning is assumed to cost $(C/2)W_0$ units of parental growth; for example, if $C = 1$, a $100\,g$ offspring is equal to $50\,g$ of forfeited maternal growth. If τ is the time required to rear one offspring to independence, then $1/\tau$ is the clutch size per unit time (or $\dfrac{1}{2\tau}$ = b since the primary sex ratio is 0.5). In τ units of time, an adult of size $W(\alpha)$ can deliver $AW(\alpha)^{0.75}\tau$ units of energy. It should be noted that the cost of an offspring is entirely given by the diversion of parental potential growth (eqn. (5.2)) to offspring. To obtain b we set $AW(\alpha)^{0.75}\tau = (C/2)W_0$, write $W_0/W(\alpha) = \delta$, and solve for $\dfrac{1}{2\tau}$.

This gives

$$b = \frac{A}{C\delta}W(\alpha)^{-0.25} \ . \tag{5.4}$$

Thus b will scale with the -0.25 power of adult body size among species with similar A, δ, and C values.

What is the value of C in eqn. (5.4)? C is between about 1.25 and 1.75, according to two independent estimates, as shown in Table 5.3.

Table 5.3. Two independent estimates of C in the offspring production relation (eqn. (5.4)). [*]

1. $S(\alpha)$ is the proportion of offspring surviving to adulthood. Millar and Zammuto (1983) show a slight positive scaling (0.08) of $S(\alpha)$ with adult body weight (actually, the inverse of reproductive value at maturation) with a central value near 0.33. By eqn. (5.9a), $C = 1.33S(\alpha)/\delta$. Since δ averages 0.33 (Fig. 5.4), we have $C = 1.3$. An intriguing possibility is that C shows a 0.08 scaling with adult body weight.

2. Figure 1.3 shows $\alpha b \approx 1.7$. By eqn. (5.9c), we have $C = (4/\alpha b)[(1 - \delta^{0.25})/\delta]$. With $\delta = 0.33$, this yields $C = 1.7$.

[*] An offspring of size W_0 at weaning is assumed to cost $(C/2)W_0$ units of parental growth. For example, if $C = 1$, a $100\,g$ offspring is equal to $50\,g$ of forfeited maternal growth.

These estimates rely on equations derived later in this chapter. The implication of $C < 2$ is that adult female mammals (including primates) are more efficient at growing babies than growing themselves. Brody (1964) summarizes the early literature on the energetic efficiency of growth and reproduction in domestic animals. His review strongly suggests that offspring production is indeed more efficient than adult growth. It would be interesting to estimate C for domestic animals using more recent and more complete data.

Natural selection on the age at maturity

The mortality rate of juveniles is generally high; suppose that it drops with age but reaches its minimum, and adult value, prior to maturation and then remains relatively constant as illustrated in Fig. 5.3. The Darwinian fitness measure R_0 can be written as

$$R_0 = \frac{b \exp\left[-\int_0^\alpha Z(x)\,dx\right]}{M}. \tag{5.5}$$

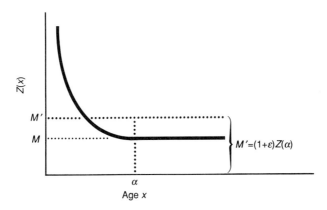

Fig. 5.3 A mortality rate assumption. $Z(x)$ is the instantaneous mortality rate at age x, and is assumed to reach a low and constant value M prior to the age α at first reproduction. Density dependence, necessary to make $R_0 = 1$, only affects $Z(x)$ for small x. Two assumptions about the adult instantaneous mortality rate are explored in this chapter. The first is that the adult mortality rate is simply M and does not increase with the onset of reproduction. The second assumption is that M increases with the onset of reproduction by a multiplier $1 + \varepsilon$. Here the adult mortality rate would be M'.

However, if the instantaneous mortality rate stabilizes prior to α, only b and $\exp\left[-\int_0^\infty Z(x)\,dx\right]$ are functions of α in eqn. (5.5). The optimal α can be found by setting $\partial\log_e R_0/\partial\alpha = 0$. We have

$$\frac{\partial\log_e b}{\partial\alpha} = Z(\alpha) \quad . \tag{5.6}$$

Since reproduction is assumed to be simply energy diverted from personal growth, $\partial\log_e b/\partial\alpha$ should be equal to $\partial\log_e (dW/dT)/\partial\alpha$ (as first noted by Kozlowski and Wiegert (1986, 1987)). Now

$$\log_e (dW/dT) = \log_e A + 0.75\log_e W;$$

thus

$$\frac{\partial\log_e (dW/dT)}{\partial\alpha} = 0.75 W^{-1}\frac{dW}{dT} = 0.75 A W(\alpha)^{-0.25} \quad .$$

This means that

$$\frac{\partial\log_e b}{\partial\alpha} = 0.75 A W(\alpha)^{-0.25} \quad .$$

If the only cost to reproduction is the growth cost implied by the diversion of resources to offspring, then $Z(\alpha) = M$, the adult mortality rate; this is assumed here, but later in this chapter a model for a mortality cost to reproduction will be developed where $M > Z(\alpha)$ as shown in Fig. 5.3. This leads directly (through eqn. (5.6)) to a scaling of the adult instantaneous mortality rate M on adult body size (for species with similar A values):

$$M = 0.75 A W(\alpha)^{-0.25} \quad . \tag{5.7}$$

As an example of the usefulness of this result consider the following. Using the estimates of A from the α allometries in Table 5.2, we predict that primates will differ in the M scaling from other more typical mammals as follows (units of kilograms and years):

$$M \approx 0.32 W^{-0.25} \qquad \text{primates, exponent may be } -0.33$$

$$M \approx 0.75 W^{-0.25} \qquad \text{other mammals} \quad .$$

Primates are thus predicted to have average adult life-spans $1/M$ about 2.5 times longer than other mammals of the same body size. The data reviewed in Table 5.2 support the predicted height for the 'other mammal' M scaling; there does not appear to be any compilation of M versus W for primates. However, indirect estimates of M, discussed later in this chapter in relation to Fig. 5.6, suggest that this predicted scaling is at least approximately correct.

It should be noted that the causal connection of M to $W(\alpha)$ is via natural selection on the age at maturity, when growth stops and all production is diverted to offspring. This argument does not have adult body size determining mortality rates; quite the opposite—it is mortality rates which determine body size. This is an answer to the question posed by Harvey and Pagel (1991) quoted at the beginning of this chapter. A special case of the relation in eqn. (5.7) (Z is a constant) has been developed by Kozlowski and Wiegert (1986, 1987).

Average immature mortality $\overline{Z}$ follows directly from $R_0 = 1$

Define $\overline{Z}$ as follows:

$$\exp\left[-\int_0^\alpha Z(x)\,dx \right] \;=\; \exp[-\overline{Z}\alpha] \quad.$$

Since $R_0 = (b/M)\exp(-\overline{Z}\alpha) = 1$ in population equilibrium, and since M, b and α are given by eqns. (5.7), (5.4), and (5.3) respectively, we can solve the relation for $\overline{Z}$ as a function of adult body size:

$$\overline{Z} \;=\; \log_e\left(\frac{1}{0.75C\delta} \right)\left(\frac{0.25A}{1 - \delta^{0.25}} \right) W(\alpha)^{-0.25} \quad. \tag{5.8}$$

Average immature mortality $\overline{Z}$ will scale with the -0.25 power of adult body size among species with similar C, δ and A values. It should be noted that M, b, and α are assumed to be independent of population density; R_0 is made equal to unity solely through density dependence on $\overline{Z}$.

5.4   Theoretical interpretations

Eliminate adult body size

We can use the growth relation (eqn. 5.3) for α to eliminate adult body size from the equations for b, M, and $\overline{Z}$, and thus derive theoretical predictions of the values of the dimensionless numbers αM, αb, and $\alpha \overline{Z}$. The proportion of offspring surviving to maturity is given by $\exp(-\alpha \overline{Z})$. The theory says that these dimensionless numbers are invariant for species with the same δ and C values. The productivity parameter A does not appear in these relations. We have

$$\alpha \overline{Z} = -\log_e(0.75C\delta) \quad \text{or} \quad \exp(-\alpha \overline{Z}) = 0.75C\delta \quad (5.9a)$$

$$\alpha M = 3(1 - \delta^{0.25}) \quad (5.9b)$$

$$\alpha b = 4\left(\frac{1 - \delta^{0.25}}{C\delta}\right). \quad (5.9c)$$

αb is the dimensionless number first introduced in Fig. 1.3, and is shown to be similar in value for primates compared with other mammals even though the α and b allometries themselves are very different for primates compared with other mammals (Fig. 1.13). Table 5.2 shows that the two groups differ greatly in the A value, while Fig. 5.4 shows that they do not differ in the δ values. Provided that the mammals have similar values of C, αb should not differ between primates and other mammals as A does not appear in these dimensionless relations.

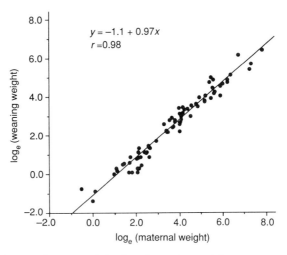

Fig. 5.4 Estimate of average δ (ratio of weaning weight to maternal weight) for 30 primates, 27 ungulates, and 20 pinnipeds. The relation is proportional with $\delta = 0.33$. Data compiled by Lee *et al.* (1991).

What if adult body size is held constant?

We rewrite eqns. (5.3), (5.4), (5.7), and (5.8) in logarithmic form:

$$\log_e \alpha = -\log_e \left(\frac{0.25A}{1 - \delta^{0.25}} \right) + 0.25 \log_e [W(\alpha)] \qquad (5.10)$$

$$\log_e M = \log_e (0.75A) - 0.25 \log_e [W(\alpha)] \qquad (5.11)$$

$$\log_e b = \log_e \left(\frac{A}{C\delta} \right) - 0.25 \log_e [W(\alpha)] \qquad (5.12)$$

$$\log_e Z = \log_e [-\log_e (0.75C\delta)] + \log_e \left(\frac{0.25A}{1 - \delta^{0.25}} \right) - 0.25 \log_e [W(\alpha)] . \qquad (5.13)$$

To understand the proposed use of these four equations, consider first eqns. (5.10) and (5.11). Suppose that a group of species with similar, but not identical, A, C, and δ values are plotted versus $\log_e [W(\alpha)]$. If the A, C and δ values do not correlate with $W(\alpha)$, then the plots of $\log_e \alpha$ and $\log_e M$ will be power functions of $W(\alpha)$, with slopes of $+0.25$ and -0.25 respectively. The intercepts will be through the average (geometric) values of $-\log_e (0.25A/(1 - \delta^{0.25}))$ and $\log_e (0.75A)$ respectively. However, individual species will fall above or below the scaling relation based upon how their particular $-\log_e (0.25A/(1 - \delta^{0.25}))$ and $\log_e (0.75A)$ values differ from the respective average values, as illustrated in Fig. 5.2. If we hold adult body size constant in eqn. (5.11), then a species i with $\log_e A_i >$ (average $\log_e A$) will have $\log_e M_i >$ (average $\log_e M$) and will fall above the fitted line. In contrast, $\log_e A_i >$ (average $\log_e A$) will cause the same species to fall, on average, below the fitted line of $\log_e \alpha$; thus a correlation calculated between $[\log_e \alpha -$ (average $\log_e \alpha$)$]$ and $[\log_e M -$ (average $\log_e M$)$]$ will be negative. Since the logarithmic scaling relations for α, M, $\overline{Z}$, and b all contain combinations of the same three parameters A, δ, and C, deviations from the average scaling relations are necessarily correlated with each other. Equations (5.10)–(5.13) can thus be used to study the correlations between $\log_e (\alpha, M, b, \overline{Z})$, expressed as deviations from their average logarithmic values, with body size held constant. As shown in Table 5.4, five of the six correlation coefficients are given unambiguously with respect to sign, while one (α, b) depends upon just why a species deviates from the average (i.e. does it differ in δ or in

Table 5.4. Theoretical signs of correlations between life history variables for mammals, with adult body weight held constant, as illustrated in Fig. 5.2.

Variable pair	Predicted correlation sign*
Adult mortality rate M Age at maturity α	−
Juvenile mortality rate $\overline{Z}$ Age at maturity	−
Juvenile mortality rate Adult mortality rate	+
Annual fecundity b Age at maturity	?†
Annual fecundity Adult mortality	+
Annual fecundity Juvenile mortality	+

* For the predicted correlation, each variable is expressed as the deviation from its mean value, as given by the theoretical scaling equations (5.10)–(5.13), with adult body weight (the $\log_e W$ term) held constant. See text for detailed development.

† This pair of equation ((5.10) and (5.12)) has two parameters in common (A, δ). Equation (5.10) contains the term $\log_e (1 - \delta^{0.25})$, while eqn. (5.12) contains $-\log_e \delta$; these two terms are *positively* related to each other. Equation (5.10) contains the term $-\log_e A$ while eqn. (5.12) contains $\log_e A$; these two are, of course, *negatively* related. Thus variation in δ generates a positive (b, α) correlation while variation in A generates a negative (b, α) correlation.

A, from the average species?). Interestingly, all four logarithmic lines share the A parameter: if variation in A is the main cause of species deviating from the average, then even the sign of α, b is given; it is negative. As shown in Table 5.4, the theoretical signs of the five correlations are exactly the same as those observed by Harvey and colleagues and listed in Table 5.1. In addition, $r(\alpha, b)$ is negative. In this exercise it is assumed that A, δ, and C are not correlated with each other.

General theoretical interpretations

The approach developed here takes A, δ, and C as given parameters and uses them to predict the other variables. Of course, we should also like to know just what determines the A, δ, and C values, as well as why the production relation is a power function with exponent c. 0.75. An evolutionary approach to the δ number is developed later in this chapter; as might be guessed, its invariance points to symmetry in the shape of a particular trade-off. Although the discussion has been couched in terms of between-species scaling, the entire formalism really refers to predictions for any combinations of A, δ and C; while the between-species scaling suggests similarities in these parameters, the real power in the approach may well lie in its freedom from any particular assumptions about them. It is probably again worth noting that while A appears in the body size relations (and may greatly influence deviation from the average scaling line), it does not appear in the αM, αb, or $\alpha \overline{Z}$ relations. The parameters δ and C are already dimensionless while A has units of $W^{0.25}$ (time)$^{-1}$; for A to remain in a dimensionless equation would require body size and age factors to cancel its units.

The equations derived here are probably the simplest which give ± 0.25 scaling and the correct correlations between residuals (Tables 5.1 and 5.4). More complex formulations might attempt to recover approximate 0.25 scaling (and the correct correlations between residuals) while allowing, for example, non-zero correlations between A, δ, and C, more complex offspring production relations, or a mortality cost to reproduction (developed later). Since the data do show 0.25 allometries, these scalings must place constraints on our model building.

5.5 One special invariant: αM (age at maturity multiplied by adult mortality)

αM is the dimensionless number introduced in Chapter 1 and studied for indeterminate growers in Chapter 4. Equation (5.9b) states that its value for a determinate grower under the present production-based theory depends only upon the relative size at independence δ. In this section we look at the expected relation $\alpha M = 3(1 - \delta^{0.25})$ in some detail for female mammals; however, first let us generalize the αM relation. Suppose that production (growth) follows the equation $dW/dT = AW^c$, where c is not necessarily 0.75. It is straightforward to solve this production relation for α and M (generalizations of eqn. (5.3) and (5.7)). This now yields the relation

$$\alpha M = \frac{c}{1 - c}(1 - \delta^{1-c}) \ , \tag{5.14}$$

i.e. we can use relations between αM and δ to ask if $c \approx 0.75$ as cxpcctcd by independently observed production relations (e.g. Lavigne 1982).

M is invcrscly proportional to α in mammals (Fig. 1.8). M and α were estimated from a sample of 26 mammal species (weaning ages in Promislow and Harvey (1990), and maturation age and life tables in Millar and Zammuto (1983); α is maturation age minus weaning age). A logarithmic plot has a slope of -0.98 (and $r = -0.88$) and the average αM equal to 0.72.

What is the average value of δ?

Millar (1977) gives relative size at weaning δ for 100 mammal species, mostly rodents with body weight less than 1 kg. The average δ was 0.37 in this sample. Charnov and Berrigan (1991*a*) estimated δ for 23 species, mostly with body weights above 1 kg (and mostly non-rodents) and found $\overline{\delta} = 0.33$. These species are shown in Fig. 5.5 and constitute a sample where the αM numbers are also estimated on a per species basis. Lee *et al.* (1991) compiled data on weaning weight and maternal weight for 30 primate species, 27 ungulate species, and 20 pinniped species. The groups have similar average values of δ (and standard errors): 0.30 (0.02), 0.34 (0.02) and 0.32 (0.03) respectively. While two of the three groups show a slight negative allometry in a plot of $\log_e W_0$ versus $\log_e W$, the pooled log–log data are almost perfectly linear with a slope of unity, as shown in Fig. 5.4. This makes the average $\delta = e^{-1.1} = 0.33$ Thus δ averages about 1/3 for mammals.

Since $1 - \delta^{0.25}$ is almost linear over the δ range 0.15–0.65, we can use $3(1 - \overline{\delta}^{-0.25})$ to predict the average αM for mammals ($\alpha M = 0.72$ in the Millar and Zammuto (1983) sample). With $\delta = 0.33$ we have $3(1 - 0.33^{0.25}) = 0.73$, virtually identical with the observed value. We obtain a very similar answer (0.75) if we average over $\delta^{0.25}$. This analysis of average αM and average δ places c in eqn. (5.14) equal to 0.75.

Is αM the same for primates?

Since $\delta \approx 0.33$ for primates also, the theory predicts that their αM should also equal 0.73. An indirect method of estimating M within primates, which uses the much more accessible data on maximum life-span in zoos (Harvey and Clutton-Brock 1985), has been developed. Figure 5.6 shows a plot of $1/M$ versus α for 15 primate sub-families (analysis at the species level gives the same relation). The fitted relation is clearly proportional, with $M^{-1} = 1.27\alpha$ or $\alpha M = 0.79$. The

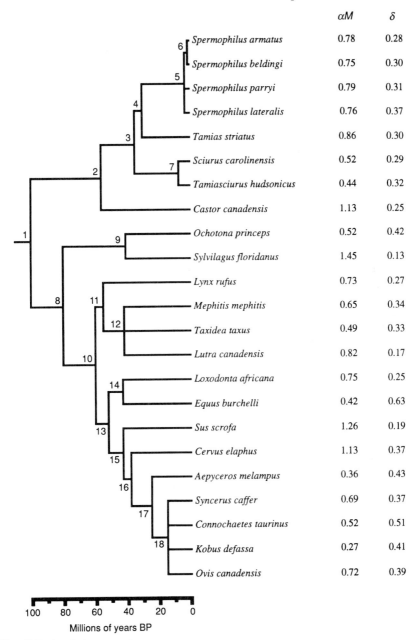

	αM	δ
Spermophilus armatus	0.78	0.28
Spermophilus beldingi	0.75	0.30
Spermophilus parryi	0.79	0.31
Spermophilus lateralis	0.76	0.37
Tamias striatus	0.86	0.30
Sciurus carolinensis	0.52	0.29
Tamiasciurus hudsonicus	0.44	0.32
Castor canadensis	1.13	0.25
Ochotona princeps	0.52	0.42
Sylvilagus floridanus	1.45	0.13
Lynx rufus	0.73	0.27
Mephitis mephitis	0.65	0.34
Taxidea taxus	0.49	0.33
Lutra canadensis	0.82	0.17
Loxodonta africana	0.75	0.25
Equus burchelli	0.42	0.63
Sus scrofa	1.26	0.19
Cervus elaphus	1.13	0.37
Aepyceros melampus	0.36	0.43
Syncerus caffer	0.69	0.37
Connochaetes taurinus	0.52	0.51
Kobus defassa	0.27	0.41
Ovis canadensis	0.72	0.39

100 80 60 40 20 0
Millions of years BP

Fig. 5.5 A phylogeny of the species in the dataset showing species values for αM and δ. Horizontal distances reflect approximate dates of branches. The numbers assigned to the nodes refer to evolutionarily independent events or contrasts. From Berrigan *et al.* (1993).

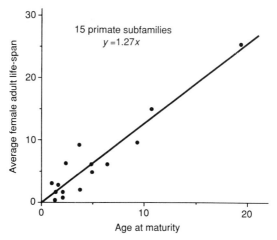

Fig. 5.6 Plot of M^{-1} versus α for 15 primate subfamilies; thus we have $(\alpha M)^{-1} = 1.27$ or $\alpha M = 0.79$, not significantly different from the values for the other mammals discussed in this chapter. Also, compare this graph with Fig. 1.8—they are the same. The fitted regression shows an intercept not different from zero and a correlation coefficient of 0.95. The standard error of the slope is 0.12. These data represent averages for a subfamily; the same analysis using species as independent data points produces the same answer. Average adult life-spans were estimated as follows. M^{-1} values for 16 non-primate species were estimated from the data of Millar and Zammuto (1983). The compilation of Eisenberg (1981) was then used to provide data on the maximum life-span T_{max} observed in zoos for the same species. A functional regression between M^{-1} and T_{max} is highly significant (units of years): $M^{-1} = 0.4T_{max} - 0.1$ ($r = 0.95$, $n = 16$). Harvey and Clutton-Brock (1985) provide primate data on T_{max} and α. This functional regression was used to estimate M^{-1} from T_{max} for the primates. This method should give a reasonable estimate of M^{-1} if the extended life-span of primates in zoos bears the same relationship to their field life-spans as is shown by other mammals.

estimation methods for M are described in the figure caption; if they are valid, αM is indeed the same for primates as for other mammals.

Does αM change with δ?

Two tests of the predicted negative relation between αM and δ are considered: the first treats the 23 species in Fig. 5.5 as independent data points (Charnov and Berrigan 1991a), while the second incorporates phylogenetic information (Berrigan *et al.* 1993).

The prediction that $\alpha M = 3(1 - \delta^{0.25})$ was tested by comparing the values of αM with the ratio δ of mass at weaning to average adult

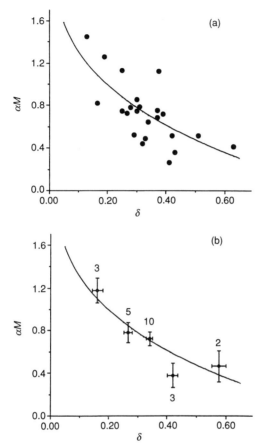

Fig. 5.7 Plots of the value of αM versus δ for females of 23 mammals species (listed in Fig. 5.5): α is the age at first reproduction adjusted for the parental care period, M is the average adult instantaneous mortality rate (Beverton 1963) and δ is the ratio of mass at weaning to adult female mass. (a) Observed ($\bullet$) and predicted (eqn. (5.9b)) (—) values of αM. The fit was tested with a linear regression of observed against predicted values ($r = 0.71$, $p < 0.001$). (b) Average of the values of δ and αM over the δ intervals 0.1–0.2, 0.2–0.3, 0.3–0.4, 0.4–0.5, and > 0.5. The error bars are one standard error, the numbers are the sample size, and the line is the predicted value ($r = 0.93$, $p < 0.03$). The statistical tests are one-tailed.

female mass for 23 species of mammals (Fig. 5.7, species listed in Fig. 5.5 and sources listed in Table 1 of Charnov and Berrigan (1991a)). The data strongly support the predicted relation. It should be noted that the r value of the linear regression between the observed and predicted values of αM increases from 0.71 to 0.93 when we fit averages (Fig. 5.7(b)) over even intervals of δ rather than all 23 points (Fig. 5.7(a)). It is not possible to distinguish between a linear regression of αM versus δ ($r = -0.67$, $p < 0.001$) and this slightly curvilinear relation ($r = 0.71$, p < 0.001). In these tests it was assumed that $c = 0.75$. The one-parameter model for c (eqn. (5.14)) was also fitted using nonlinear regression. The solution from fitting all 23 points, plus or minus one standard error, gives $c = 0.74 \pm 0.03$ ($r = -0.71$, $p < 0.001$) and the solution for the five average values gives $c = 0.75 \pm 0.04$ ($r = -0.93$, $p < 0.03$). Because the production relation of eqn. (5.2) appears to be an important component of mammalian life histories, it is particularly significant that the exponent 0.75 appears in the relation between αM and δ at evolutionary equilibrium. This analysis is the first indirect determination of this exponent and gives virtually the same answer as direct measurements of individual or offspring production.

Phylogenetic contrasts and αM

Berrigan *et al.* (1993) rejected the normalizing selection assumption and suggested a historically based test of the $\alpha M - \delta$ hypothesis. First, using literature sources, they constructed a phylogeny for the 23 species in the data set (Fig. 5.5). They then applied Felsenstein's (1985) method of independent comparisons modified to deal with multiple nodes (Harvey and Pagel 1991, p. 157) to the 23 species and generated 18 sets of independent contrasts (Fig. 5.5). These contrasts can be thought of as representing comparisons between species at the tips of branches or between nodes within a tree. The degrees of freedom available for the comparison are adjusted to reflect the number of independent evolutionary events which are being examined. They are asking if αM and δ change together in evolutionary time and if the changes cluster around the line $\alpha M = 3(1 - \delta^{0.25})$. Over the range of δ in the data set the equation $\alpha M = -5.17 (\log_e \delta^{0.25})$ is a good linear approximation to the predicted relation. They tested the hypothesis that the slope relating contrasts in αM to contrasts in $\delta^{0.25}$ was equal to -5.17, as determined by linear regression through the origin. They estimated the error in x versus y, and showed that ordinary regression was (approximately) correct as $\delta^{0.25}$ had a much smaller error than αM. The estimated slope was -6.35 (95 per cent confidence intervals, -9.7 to -3.0), not far from the predicted value of -5.17. With normalizing selection each species is an independent evolutionary event; under this directional selection assumption, the shifts in values are the independent events. The above analysis shows that the shifts in δ and αM are along the theoretical $\alpha M - \delta$ line.

What determines δ?

The ecological correlates of differences in δ within the mammals are not obvious. The three largest values of δ in this study (Fig. 5.5) are those for the impala, wildebeest, and zebra, and the three lowest are those for the rabbit, otter, and boar; squirrels and elephants have similar intermediate values of δ (and αM). A synthetic approach to the evolution of δ is developed here.

Harvey and Nee (1991) pointed out that eqn. (5.7) implies that adult body size (and M) is independent of the size at weaning W_0; thus, with respect to changes in W_0, fitness for a mutant individual ($\tilde{R}_0$ in eqn. (5.5)) can be written as $\log_e \tilde{R}_0 = \log_e \tilde{b} + \log_e \tilde{S}(\alpha) - \log_e M$. The equilibrium W_0 will occur where

$$\frac{\partial \log_e \tilde{R}_0}{\partial \tilde{W}_0} = 0 \quad \text{or} \quad \frac{\partial \log_e \tilde{b}}{\partial \tilde{W}_0} = -\frac{\partial \log \tilde{S}(\alpha)}{\partial \tilde{W}_0} \quad (5.15)$$

when the " ˜ " variables are set equal to the population values. According to the model for b (eqn. (5.4)) $\partial \log_e \tilde{b}/\partial \tilde{W}_0 = -1/\tilde{W}_0$. $\tilde{W}_0$ may affect $\tilde{S}(\alpha)$ through altered time under maternal protection, an increased survival rate after independence, and/or shortened time to age α (size $W(\alpha)$).

Suppose that for $\tilde{S}$ and $\tilde{W}_0$ near the population values (S, W_0) we can write

$$\tilde{S} = S\left[1 + G\left(\frac{\tilde{W}_0 - W_0}{W(\alpha)}\right)\right] .$$

This is a linear approximation of the $\tilde{S}, \tilde{W}_0$ function which is valid near S and W_0; G is the slope (the shape) of this function. Now,

$$\log_e \tilde{S} = \log_e S + \log_e\left[1 + G\left(\frac{\tilde{W}_0 - W_0}{W(\alpha)}\right)\right]$$

and

$$\frac{\partial \log_e \tilde{S}}{\partial \tilde{W}_0} = \frac{G/W(\alpha)}{1 + G[(\tilde{W}_0 - W_0)/W(\alpha)]} .$$

Setting $\tilde{W}_0 = W_0$ in this function yields

$$\frac{\partial \log_e \tilde{S}}{\partial \tilde{W}_0} = G/W(\alpha) .$$

Equation (5.15) gives the ESS, or $-G/W(\alpha) = -1/W_0$. Finally, $W_0/W(\alpha) = 1/G$. $W_0/W(\alpha)$ is δ, and so the final result is $\delta = 1/G$. All species with the same δ are predicted to share the trade-off shape as given by G.

Of course, this model for the evolution of δ assigns control to the mother as opposed to the offspring (Trivers 1974, Godfray and Parker 1991). There do not appear to be any data which provide independent estimates of the shape parameter G.

5.6 A mortality cost of reproduction?

Estimation of various costs of reproduction has become an immense growth area within life history studies (Shine 1980; Reznick 1985; Bell and Koufopanou 1986; Clutton-Brock 1991; Lessells 1991; Partridge and Sibly 1991; Shine and Schwarzkopf 1992; Stearns 1992). Much of the literature focuses on the demonstration *per se* of a trade-off between present and future reproduction—a mortality cost and/or a future fecundity cost. The results have been quite mixed (some studies show a cost but many do not, particularly not a mortality cost) with much discussion devoted to the methodological issues of just how the presence or absence of a reproductive cost can be inferred. The literature is populated with arguments about the merits (and pitfalls) of using genetic correlations, phenotypic correlations, or experimental manipulations for such inference.

The theory developed for mammals formerly allowed only a future fecundity cost through the curtailment of personal growth. The theory is altered here by adding a mortality cost to reproduction. There are, of course, a number of ways that this could be done (e.g. is the cost size dependent or age dependent?). However, it seems reasonable to require that the expanded theory also predicts 0.25 allometries and the correct correlation structure among the residuals; after all, these are the patterns in the data. There are several ways of adding the mortality cost but most destroy the 0.25 allometric structure. The method reported here is the simplest addition of mortality cost which also gives 0.25 allometry and the correct correlations among residuals. This alteration of the model structure immediately suggests a way of testing for the relative importance of the mortality cost since it is the heights (intercepts) of the mortality rate allometries M and $\overline{Z}$ that are changed.

In the earlier theory the average adult instantaneous mortality rate M was the same as the non-reproductive mortality rate $Z(\alpha)$ at age α. We now add a mortality cost to reproduction by making $M = Z(\alpha)(1 + \varepsilon)$ where $\varepsilon \geq 0$ (M' in Fig. 5.3). For example, if $\varepsilon = 0.5$, reproduction increases the instantaneous mortality rate by a multiplier equal to 1.5. ε is assumed to be independent of age but is not necessarily the same for all species or populations.

In equilibrium α is given by the same growth equation (eqn. (5.3)):

$$\alpha = \frac{4(1 - \delta^{0.25})}{A} W(\alpha)^{0.25}$$

However, M takes a slightly different form. The steps leading to eqn. (5.7) yield the intermediate result

$$Z(\alpha) = 0.75AW(\alpha)^{-0.25}$$

If $M = Z(\alpha)(1 + \varepsilon)$, we have

$$M = 0.75(1 + \varepsilon)AW(\alpha)^{-0.25} \quad .$$

Forming the product of αM gives

$$\alpha M = 3(1 + \varepsilon)(1 - \delta^{0.25}) \quad .$$

However, we already know that αM is fitted well by the relation $3(1 - \delta^{0.25})$ (see Fig. 5.7 and text discussion). In Fig. 5.8 αM is plotted versus $1 - \delta^{0.25}$ and the line for $\varepsilon = 0.2$, a 20 per cent mortality cost to reproduction, is drawn to show that almost all the data fall below this line, indicating no support for even a 20 per cent mortality cost to reproduction. The comparative data place $\varepsilon \approx 0$; the broad patterns of mammalian life history are consistent with a zero mortality cost to reproduction. While it cannot be shown that the above method is the only way of adding a mortality cost which preserves the 0.25 allometry and correct residual correlations, this comparative test placing $\varepsilon \approx 0$ may provoke others to study the problem further.

5.7 Sexual dimorphism in adult body size

If the ratio

$$\frac{\text{adult body weight for sex } i}{\text{adult body weight for sex } j}$$

depends upon which sex (male or female) is i and which is j, we have an asymmetry.

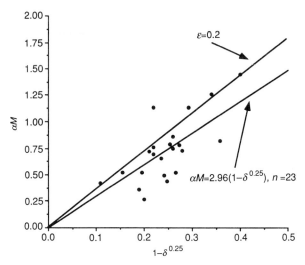

Fig. 5.8 Regression of αM on $1 - \delta^{0.25}$ with the model $y = Dz$; the estimated slope is 2.96 (95 per cent confidence intervals, 2.61–3.3), not different from the value of 3 obtained with the no-mortality cost ($\varepsilon = 0$) model. Interestingly, the regression of $1 - \delta^{0.25}$ on αM (model is $y = z/D$) yields a slope D of 3.16, again not different from 3. $\varepsilon = 0.2$ refers to a 20 per cent mortality cost to reproduction. Almost all the data fall below the $\varepsilon = 0.2$ line, indicating no support for even a 20 per cent mortality cost to reproduction.

Asymmetry under this transformation is sexual dimorphism in body size and has been of great interest to evolutionists since Darwin developed his theory of sexual selection. Explanations of sexual dimorphism in adult body size have usually focused on reproductive and/or ecological performance as a function of gender and body size (Shine 1978, 1979, 1988, 1989, 1990; Alexander *et al.* 1979; Berry and Shine 1980; Arak 1988; Reiss 1989). However, even under determinate growth the problem is made more complex by the fact that adult body size is the result of growth, which takes time. Therefore the evolution of adult body size is tied to evolution of the age at maturity, when growth stops and the production of offspring starts (see also Shine 1990; Parker 1992). With indeterminate growers, the problem is even more complicated since the concept of adult size requires decisions about how to average over adult sizes.

In this section the life history theory developed in this chapter is extended to male mammals, and allows us to predict both adult male body size and, by division, sexual dimorphism. The argument leading to eqn. (5.7) for the equilibrium size at maturity assigned to females a fecundity per unit time equal to the individual's production rate, which scaled with body size W_f to the power 0.75. For males, let us assume an arbitrary power, or that fecundity per unit time scales with W_m^y. The exponent y can thus be used to scale male reproductive ability to male body size; large y means that large size is quite important in male reproductive success, whereas small y assigns less reproductive importance to size.

In the earlier model, post-weaning (but pre-reproductive) growth follows the growth law $dW/dT = AW^{0.75}$. Here A is simply assigned a sex-specific subscript (i.e. A_m for males and A_f for females). Males may conceivably also experience different immature mortality rates. Let the instantaneous mortality rates near the ages of maturity be M_m for males and M_f for females. The average lengths of the adult life-span for females and males are $1/M_f$ and $1/M_m$ respectively; denote these as E_f and E_m. It is simple to extend this to include a mortality cost to reproduction.

With these definitions we can now use the life history model to solve for the ESS body size in males W_m and form the ratio W_m/W_f to obtain a measure of sexual dimorphism. The size at maturity for males and females respectively follows from the argument for eqn. (5.7):

$$W_m^{-0.25} A_m y = M_m$$

$$W_f^{-0.25} A_f \times 0.75 = M_f .$$

We can combine these to yield

$$\log_e \left(\frac{W_m}{W_f} \right) = 4\log_e \left(\frac{E_m}{E_f} \right) + 4\log_e \frac{y}{0.75} + 4\log_e \left(\frac{A_m}{A_f} \right) . \tag{5.16}$$

Equation (5.16) can be interpreted as follows. Dimorphism is seen in evolutionary equilibrium to be equal to the sum of three parts, each of which summarizes a growth, reproductive, or survival difference between the sexes. These three dimensionless numbers combine equally in a sum rule to generate the dimorphism theoretically. E_m/E_f is the ratio of male to female average adult life-span, A_m/A_f is the ratio of male to female growth efficiency (i.e. height of the dW/dT curves), and $y/0.75$ is the ratio of the exponents (or powers) which translate adult body weight into the fertility rate (y for males, 0.75 for females). This last ratio is a comparison of the implications of body size for male and female reproductive ability. Female production (the 0.75 scaling) is really female–female competition for resources for offspring; likewise, y gives male–male competition for the production of offspring. Thus their ratio computes a between-sex comparison of the importance of body size to reproduction.

Sum rules like eqn. (5.16) are particularly useful in that they show how the dimorphism potentially results from the interplay of various life history factors; we can then ask if one or another factor is important in particular ecological (or historical) settings, or we can ask how large a single factor must be if it alone causes the dimorphism. For example, males and females in the northern elephant seal (*Mirounga angustirostris*) have nearly identical adult life-spans, but the W_m/W_f ratio is approximately 12:1, making it the most sexually dimorphic mammal (Deutsch *et al.* 1990; Reiter and LeBoef 1991). Provided that $A_m \approx A_f$, we can solve eqn. (5.16) to find $y = 1.4$. Thus males are predicted to be gaining fertility with size at about twice the rate of females (1.4 versus 0.75). The most surprising implication of this calculation is how small a dimorphism in, say, size–fertility relations is needed to produce a ten-fold body size dimorphism; indeed, a twofold body size dimorphism requires a y of only 0.89, only about 20 per cent larger than the female value. Can field data really demonstrate a 20 per cent difference? However, there is at least one indirect way of estimating y. Under the growth model assumed here (eqns. (5.2) and (5.3)), αM for males takes the following form: $\alpha M = 4y(1 - \delta_m^{0.25})$, where δ_m is the ratio of male weight at weaning divided by adult male weight. While αM will rarely be known with high precision, comparative data for several species with similar W_m/W_f ratios may allow a sensible regression estimate of y, particularly if a mortality cost to reproduction can be ruled out.

5.8 Summary

Any theory for the comparative structure of female mammalian life histories must produce at least approximate 0.25 allometries for b, M, $\overline{Z}$, and α; in doing so, such theories will also predict what may cause species to deviate from the fitted allometry, the correlations between residuals, and the numerical values for αM, αb, and $\alpha \overline{Z}$. In this chapter a production–growth relation of the form $dW/dT = AW^{0.75}$ is used in the face of externally imposed mortality to implement this pro-gramme. The production relation alone gives the body size scalings of age at first reproduction α and yearly fecundity b. Natural selection on the age at maturity produces the adult mortality M scaling, while the assumption of a population held stationary ($R_0 = 1$) through density-dependent juvenile survival yields the scaling for immature mortality $\overline{Z}$. The present theory correctly predicts the 0.25 allometries and the signs of the correlations between their residuals (Fig. 5.2, Tables 5.1 and 5.4).

The predicted heights of the four allometries are greatly affected by the height A of the production relation. However, the three dimensionless numbers αM, αb, and $S(\alpha) = \exp(-\alpha \overline{Z})$ are independent of A; they are only functions of two other dimensionless numbers, the relative size δ of an offspring at weaning and the efficiency C of a mother growing an offspring (relative to growing herself). Predictions about αb and αM are supported by a comparison of primates ($A \approx 0.40$) versus other mammals ($A \approx 1$ on average). The primate allometries for α and b are very different from those of most other mammals (Fig. 1.13), but the αb number is the same (Fig. 1.3). Primates have the same average δ as other mammals (approximately 0.33), and so we predict they will have the same average αM value (approximately 0.73); preliminary calculations suggest that this is true (Fig. 5.6). Data from a wide variety of mammals also support the predicted negative relation between αM and δ (Fig. 5.7). Finally, the theory is extended to include a mortality cost to reproduction; the comparative data place this potential cost near zero (Fig. 5.8). The chapter ends with a discussion of sexual dimorphism in adult body size.

I consider the theory developed in this chapter to be a hopeful beginning—hopeful because the comparative data do support a number of the predictions, a beginning because so few mammals have been included in the tests and because there are other life history variables that have not yet been examined (e.g. components of 'yearly clutch size' b). Can it really be true that most variation in the allometric structure of mammalian life histories traces to variation in the height A of a basic production relation $dW/dT = AW^{0.75}$? If so, *why* do primates have such low A values? Labelling the low primate value of A a phylogenetic constraint only begs the question of what determines its value.

6

Population dynamics

6.1 Introduction

In this chapter we consider some general patterns in the dynamics of populations and ask what symmetry or invariance for population size and body size transitions can tell us about them, particularly for organisms with determinate growth. The population size invariants examined are the age at first reproduction α and the adult instantaneous mortality rate M. Of course, a body size invariant is something that does not change with body size ($\propto W^0$); the one examined here is the relationship between the net reproductive rate R_0 and the relative population size. Our old friends αM and αb will also put in an appearance. The population dynamic patterns to be explored are as follows (Charnov 1992): (1) the -0.25 scaling of the maximum intrinsic rate of increase $r_{\max}$ with adult body weight W, (2) Fowler's (1988) rules which link the relative position of the inflection point of the population growth curve (relative to the carrying capacity K) to the maximum net reproductive rate R_{0m} (R_0 for an individual in a rarified population).

6.2 r_{max} allometry

The intrinsic rate of increase r is given by the relation

$$\frac{\mathrm{d}\log_e N}{\mathrm{d}T} = r \quad \text{or} \quad \frac{\mathrm{d}N}{\mathrm{d}T} = rN \ . \tag{6.1}$$

The maximum intrinsic rate of increase r_m or $r_{\max}$ is a concept defined in at least two distinct ways in the literature. First, as used by many biologists who study organisms of insect size or smaller, it refers to a maximum as expressed under approximately ideal growth conditions in the laboratory. Such estimates are useful for considering, for example, potential pest or disease outbreaks (Anderson and May 1991) or plankton blooms, which is why numerous data have been gathered for organisms such as spider mites, algae, etc. The second concept of r_{max} is the one which results from the following thought experiment. The population size N is greatly reduced, or rarified, from its typical size K, and then its intrinsic rate of increase r is observed. r should be at its maximum, for field conditions, in this rarified population. This second r_{max} is really the one that appears in ecology textbooks in a great many equations for population dynamics. For this case we depress the population from the carrying capacity K but hold constant the rest of its ecological relations, as expressed in its usual habitat. In general, r_{max} for this case should be less than r_{max} under ideal growth conditions in the laboratory.

The data

Fenchel (1974) first showed, primarily under laboratory conditions, that r_{max} gives a -0.25 scaling with adult body weight. He claimed that there were really three separate curves, each with slope -0.25 but with different heights (intercepts): from high to low he had homeotherms (endotherms), poikilotherms (ectotherms), and unicells. He noted that the three lines were ordered in height in the same way as the -0.25 scalings in metabolic rate per unit body weight for the three groups (e.g. Fig. 5.1 of Stearns (1992)). He also pointed out that the individual growth rates of the homeotherms were much higher than the others, resulting in a larger body size for a fixed age at maturation (see Fig. 3 of Fenchel (1974)). From this he concluded that the r_{max} scaling ought to be higher for the homeotherms. Southwood (1981) extended Fenchel's argument by using an approximate equation for r_{max} (eqn. (6.5) here) and suggested that the linkage to relative metabolic rate lay in the known negative correlation of it to longevity. Blueweiss *et al.* (1978) combined Fenchel's data with data from several other sources; they found no statistical difference in height between the three curves and pooled the data (plus other data) to produce Fig. 6.1. However, since the three groups showed almost no overlap in body size, it is not too surprising that little statistical power existed to show differences in intercept. Their units are grams and days; here, the equation is also given for units of kilograms and years. They ignored Fenchel's argument as to why the lines ought to differ.

Figure 6.2 shows $\log_e r_{max}$ derived from field data versus weight for eight herbivorous mammals (Caughley and Krebs 1983). It should be noted that all of these would occupy only the extreme right-hand corner

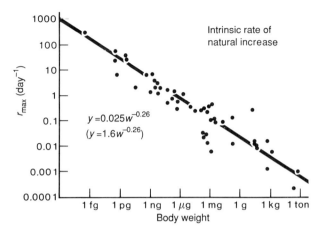

Fig. 6.1 The intrinsic rate of population growth r_{max} (day^{-1}) versus adult body size W(g) for a broad range of organisms. Data compiled by Blueweiss *et al.* 1978 (see for additional sources). Most are laboratory data; not all are reproduced here. The equation in parentheses is for units of kilograms and years.

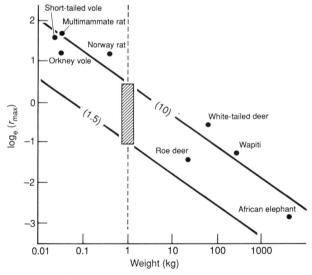

Fig. 6.2 r_{max} (year^{-1}) versus W (kg) for field populations of eight herbivorous mammals. The weight is the average of females and males. Lines indicate theoretical predicted relations (eqn. (6.10)) for $A_1 = 0.75$, $A_2 = 1$, and R_{0m} values 1.5 and 10 respectively; relations for $1.5 < R_{0m} < 10$ fall between these two extremes. R_{0m} is the net reproductive rate R_0 in a rarified population. Data compiled by Caughley and Krebs (1983).

of Fig. 6.1. There are really only two clusters of points on Fig. 6.2, and so the regression estimate of the scaling exponent (-0.36, discussed further below) is not very informative even if the correlation is very high (coefficient of determination is 0.92). The weights are also means of males and females.

Although there are some other data sets in the literature (Peters 1983; Hennemann 1983; Calder 1984; Gaston 1988; Ross 1992), those discussed above suffice for present purposes; the Ross and Hennemann data will be considered later. To resolve disputes about r_{max} allometry we need to specify which concept of r_{max} is being studied and we also need to develop theory specifically to predict the intercepts and the slopes of the scaling lines. This chapter develops two approaches to this, building on the earlier suggestions of Fenchel (1974) and Southwood (1981), particularly Southwood's use of an approximate model for r_{max} (eqn. (6.5)).

Why r_{max} allometry?

Consider organisms with the following life history, which closely approximates mammals. Growth in body size is determinate and ceases at the age at maturity α. Adulthood has constant yearly fecundity b (in daughters) and constant mortality (the adult instantaneous mortality rate is M). The proportion of offspring who survive to maturity, age α, is $S(\alpha)$. The net reproductive rate R_0 and the population's intrinsic rate of increase r are given by the usual demographic equations (e.g. Pianka 1988):

$$R_0 = \frac{bS(\alpha)}{M} \tag{6.2}$$

$$1 = \int_0^\infty L(x)b(x)\exp(-rx)\,dx \quad .$$

With $b(x) = 0$ for $x < \alpha$ and $L(x) = S(\alpha)\exp[-M(x - \alpha)]$ for $x > \alpha$, this can be written as

$$1 = \frac{bS(\alpha)}{\exp(r\alpha)} \int_\alpha^\infty \exp[-(M + r)(x - \alpha)]\,dx \quad . \tag{6.3}$$

Evaluating the integral in eqn. (6.3) yields

$$1 = \frac{bS(\alpha)}{\exp(r\alpha)} \frac{1}{M + r}$$

or

$$\exp(r\alpha)(M + r) - bS(\alpha) = 0 \ . \tag{6.4a}$$

This can be combined with eqn. (6.2) and rewritten as

$$\exp(r\alpha)(M + r) - MR_0 = 0 \ . \tag{6.4b}$$

There is also a well-known approximate model for r (Southwood 1981; Fowler 1988; Pianka 1988), namely

$$r \approx \frac{\log_e R_0}{T_c} \tag{6.5}$$

where T_c is the average generation length defined as

$$T_c = \frac{\displaystyle\int_x xL(x)b(x)\,\mathrm{d}x}{\displaystyle\int_x L(x)b(x)\,\mathrm{d}x} \ .$$

The approximation is good if R_0 is small or if the coefficient of variation in T_c is not too large (even if R_0 is large (May 1976)). For the determinate growth life history (eqn. (6.3)) $T_c = \alpha + 1/M$, the average age of adult female, and so we have

$$r \approx \frac{\log_e R_0}{\alpha + 1/M} \ . \tag{6.6}$$

We can also use eqn. (6.6) as an approximation for indeterminate growers. This result (eqns. (6.5) and (6.6)) means that $r \propto T_c^{-1}$ for species with similar R_0 values, which has been discussed by Southwood (1981), Heron (1972), and, originally, Smith (1954).

I fully realize that the intrinsic rate of increase given by eqns. (6.4)–(6.6) is the idealized rate for a population assumed in stable age distribution; however, it is the only rate that we can calculate independent of detailed specification of existing age distributions, and so it is used here.

Now consider the following thought experiment for a determinate grower. The population size is greatly reduced (rarified) from its typical size, and its intrinsic rate of increase r is then observed; r should be at its maximum in this rarified population. Equation (6.4b) gives r, here treated as r_{max}, and allows us to ask for conditions on M, α, and R_0 which will force

$$r_{max} = A_3 W^{-0.25} \qquad (6.7)$$

for a collection of species with different body weights W but the same A_3 value. A_3 is simply the height of the r_{max} scaling line.

Suppose that we choose a species and write M and α for its rarified population as follows:

$$M = A_1 W^{-0.25} \qquad (6.8)$$

$$\alpha = A_2 W^{0.25} \quad . \qquad (6.9)$$

If M and α are density-independent parameters, as discussed in Chapter 5, A_1 and A_2 are given by the life history equations (5.7) and (5.3) respectively ($A_1 = 0.75A$ and $A_2 = 4(1 - \delta^{0.25})/A$).

If we substitute eqns. (6.8) and (6.9) into (6.4b), we force r to become a function of W and R_0. For a fixed R_0 there will be only one function which relates r to W, and that function is the $A_3 W^{-0.25}$. This is easily seen by substituting (6.7), (6.8), and (6.9) into (6.4b) and noting that the following factorization is allowed:

$$W^{-0.25}[\exp(A_2 A_3)(A_3 + A_1) - A_1 R_0] = 0 \quad . \qquad (6.10)$$

A group of species with differing values of W but the same A_1, A_2, and R_0 in the rarified condition (R_{0m}) will share the same A_3; they will satisfy $r_{max} = A_3 W^{-0.25}$. In Chapter 5 A_1 and A_2 were derived for mammals (eqns. (5.7) and (5.3)), making A_1 and A_2 independent of population size; all density dependence was assumed to affect immature survival $S(\alpha)$. The net reproductive rate in a rarified population R_{0m} is greater than unity solely through the increase in $S(\alpha)$. Since mammals share approximately the same A_1 and A_2 values, the condition for r_{max} to scale as $W^{-0.25}$ then reduces to the various species having similar values of R_0 in rarified populations (Fowler (1988) also recognizes the importance of this body size invariance for R_{0m}). If, when rarified, the species differ slightly in A_1, A_2, or R_0, they may still approximate

-0.25 scaling for r_{max} if these parameters are not correlated with body size. Equation (6.10) can be used to predict whether a species will fall above (A_3 high) or below (A_3 low) the average r_{max} scaling line.

How high is the mammal r_{max} line?

With $A = 1$ and $\delta = 0.33$ for a typical mammal (Table 5.2) we have $A_1 \approx 0.75$ and $A_2 \approx 1$ (units are years and kilograms). Figure 6.3 shows A_3 versus R_{0m} for these values inserted into eqn. (6.10) (a very similar answer would follow from eqn. (6.6)). Recall that $A_3 = r_{max}$ for a 1 kg animal. The -0.25 scalings for $R_{0m} = 1.5$ and $R_{0m} = 10$ are plotted in Fig. 6.2; these bound A_3 for the intermediate R_{0m} values. It is not possible to claim any sort of quantitative fit (r_{max} reported for elephants is about three times too large given the R_{0m} reported by Fowler (1988) (Table 6.1)), but clearly the mammals fall in the neighbourhood of the predicted relations (and perhaps R_{0m} is near 10 for these rodents). That Caughley and Krebs (1983) found a negative slope of 0.36 instead of 0.25 probably follows from the fact that the small rodents have higher values of R_{0m} than the elephant (Table 6.1). A theoretical relation like eqn. (6.10) that predicts the height of a scaling line may stimulate field workers to generate more precise r_{max}, R_{0m}, and body size data.

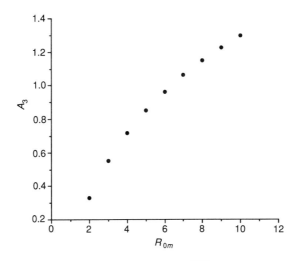

Fig. 6.3 Height of the scaling $r_{max} = A_3 W^{-0.25}$ as a function of R_{0m} for $A_2 = 1$ and $A_1 = 0.75$ (the mammal values from Table 5.2 since $A = 1$) (from eqn. (6.10)).

Table 6.1. Species used to calibrate Fowler's rule: relative population size ($\theta = N/K$) at the inflection point for population growth (θ^* at max dN/dT) is inversely related to the net reproductive rate in a rarified population ($R_0 \rightarrow R_{0m}$ as $\theta \rightarrow 0$).

Species	$\log_e (\log_e R_{0m})$	R_{0m}	θ^* (in rank order)
Elephant	− 0.94	1.48	0.86
Sheep	− 0.76	1.60	0.85
Stenella	− 0.93	1.48	0.76
Fin whale	− 0.47	1.87	0.81
Escherichia	− 0.37	2.00	0.67
Paramecium	− 0.37	2.00	0.61
Fur seal	− 0.19	2.29	0.60
Grizzly bear	0.20	3.39	0.66
Bighorn sheep	− 0.01	2.69	0.62
Deer	0.37	4.25	0.56
Grey whale	− 0.05	2.59	0.53
Blue whale	0.42	4.58	0.59
Mouse	0.55	5.66	0.50
Striped bass	0.85	10.38	0.50
Daphnia	1.49	84.5	0.37
Drosophila	1.53	101.3	0.37
Tribolium	2.06	2555	0.26

Data from Fowler, personal communication.

How high is the ectotherm r max line?

For a typical mammal with $R_{0m} \approx 2.5$–4, Fig. 6.3 yields the scaling $r_{max} \approx 0.6 W^{-0.25}$. Suppose that we treat the ectotherms as having determinate growth and following the same production relation ($dW/dT = A W^{0.75}$) as the mammals. What is the value of A for the ectotherm? Homeothermic individuals use energy at a rate about 30 times that of an ectotherm of the same body size (Peters 1983; Damuth 1987); their production (growth) rate may be at least 5–10 times larger (Fenchel 1974; Case 1978; Stearns 1992, Fig. 5.2). This makes the value of A for the ectotherm in the production equation about 0.1–0.2, in contrast with $A \approx 1$ for the mammals. Using the mammal life history equations for α and M (5.3 and 5.7 with $\delta \approx 0$) to give us A_1 and A_2, setting $R_{0m} = 10$ (a fish, Table 6.1) and $A = 0.2$, and substituting these into (eqn. (6.6))

$$r_{max} \approx \frac{\log_e R_{0m}}{A_2 + 1/A_1} W^{-0.25},$$

we finally obtain $r_{max} \approx 0.075 W^{-0.25}$ for the ectotherms.

The field-derived endotherm scaling line ought to be about a factor of $0.6/0.075 = 8$ higher than the ectotherm line. An ectotherm with this growth potential ($A \approx 0.2$) would require R_{0m} near 100 to be on the same scaling line as the endotherm. The value of these calculations lies in the attempt to predict A_3 from more basic considerations of the life histories; the slower individual growth rate (and not the lower mass-specific metabolic rate itself) of an ectotherm almost certainly implies a much lower A_3 value, as Fenchel (1974) originally argued. This conclusion is not strongly dependent upon the life history model assumed above but follows mostly from the much slower individual growth rate for the ectotherm.

Mammals once again: primates versus others

Like ectotherms, primates have a relatively low A value ($c.$ 0.4, Table 5.2); thus we predict that they will have a relatively low A_3 value. Figure 6.4 shows $\log_e r_{max}$ versus $\log_e$ (body weight) for a sample of

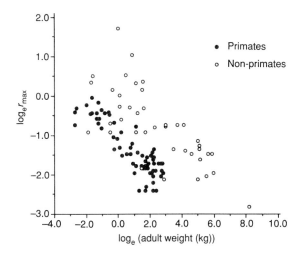

Fig. 6.4 r_{max} versus adult body weight for 72 primate species (Ross 1992) and 40 species of other mammals (Hennemann 1983). As expected from their slow personal growth rates ($A \approx 0.4$ (primates) versus $A \approx 1.0$ (others)), the r_{max} line is much lower for the primates. The estimated $\log_e$ intercepts are 0 and -1.1 respectively (standard error, 0.06); analysis of covariance shows a common slope of -0.31 (standard error, 0.023). The coefficient of determination r^2 is 0.67. r_{max} was estimated the same way in both these studies using an equation derived by Cole (1954) for r. Cole's equation is (approximately) the same as eqn. (6.4a) with the added assumption of no mortality ($M = 0$, $S(\alpha) = 1$). See text for further discussion.

72 species of primates (Ross 1992) and 40 species of other mammals (Hennemann 1983). These two studies used the same estimation procedure (detailed below) for r_{max}, which is why they are plotted on the same graph. Analysis of covariance shows the same slope (-0.31; standard error, 0.023) but different heights for the two data sets. We have the following fitted lines:

$$r_{max} \approx 1W^{-0.31} \qquad \text{(other mammals)}$$

$$r_{max} \approx 0.33W^{-0.31} \qquad \text{(primates)} \quad .$$

Ross and Hennemann estimated the r_{max} values using an equation originally derived by Cole (1954). Cole's approach to the intrinsic rate of increase basically consists of a discrete-time version of eqn. (6.4a) with the additional assumption of zero mortality ($M = 0$, $S(\alpha) = 1$). We thus have from eqn. (6.4a)

$$r \exp(r\alpha) - b = 0$$

or

$$\alpha r \exp(r\alpha) = \alpha b \quad .$$

The two studies estimated α and b for the various species and used this equation to extract r (actually, Cole's equation is slightly different from this but the difference is unimportant for our discussion).

This equation is interesting in that species with the same αb value are predicted to share the same αr value. From Fig. 1.3 we know that $\alpha b \approx 1.7$ for primates *and* other mammals. This makes $\alpha r = 0.78$ or $r_{max} = 0.78/\alpha$ for both groups. From Table 5.2 we have

$$\alpha \approx 1W^{0.25} \qquad \text{(others)}$$

$$\alpha \approx 2.4W^{0.34} \qquad \text{(primates)} \quad .$$

These yield heights for the r_{max} lines of 0.78 (others) and 0.33 (primates), close to the fitted values (1, 0.33). This means that the difference between primates and others on Fig. 6.4 is almost entirely due to the difference in their respective α allometries.

While it may not be too unrealistic to set $S(\alpha) \approx 1$ for a rarified population, it is clearly not reasonable to take M to be zero. For $M \neq 0$ (but $S(\alpha) = 1$) we can rewrite eqn. (6.4a) as

$$\exp(r\alpha)(\alpha M + r\alpha) = \alpha b$$

Species with the same αM *and* αb will now have constant $r\alpha$. If $\alpha M = 0.75$ and $\alpha b = 1.7$ for all mammals (on average), we obtain $r_{max} = 0.4/\alpha$. This yields predicted scalings of

$$r_{max} \approx 0.17 W^{-0.34} \qquad \text{(primates)}$$

$$r_{max} \approx 0.4 W^{-0.25} \qquad \text{(others)} \quad .$$

The height of the 'other' line is 0.4 rather than the 0.6 given previously (see p.121) for mammals because $\alpha M = 0.75$ and $\alpha b = 1.7$ correspond to $R_{0m} \approx 2.3$ with $S(\alpha) = 1$. (On p.121 we set $R_{0m} \approx 2.5$–4.)

6.3 Fowler's rules

In a characteristically original paper, Fowler (1988) has produced data in support of two population dynamic rules which are independent of body size for organisms as varied as *Tribolium*, *Escherichia coli*, and many mammals. Let K be the carrying capacity and set $\theta = N/K$, the population size N relative to K; let R_{0m} be R_0 for the rarified condition. Finally, define θ^* as the value of θ which results in the maximum population growth rate (largest dN/dT, the inflection point of the population growth curve). This is illustrated in Fig. 6.5. Fowler (1988) then shows that (1) species with the same R_{0m} share the same θ^* and (2) θ^* is inversely related to R_{0m} (his data fit $\theta^* = 0.633 - 0.187y$ where $y = \log_e(\log_e R_{0m})$). Table 6.1 summarizes the data and Fig. 6.6 shows the remarkably good linear relation. This relation is independent of body size, with whales and *Escherichia coli* close together. Since R_{0m} plays such an important role in r_{max} allometry, it is interesting to see (Table 6.1) that it is not related to body size *per se*.

In this section we use the approximate r equation (eqn. (6.6)) to discuss Fowler's rules (see Charnov (1992) for an approach using the more exact relation (eqn. (6.10)). Equation (6.6) can be rewritten in dimensionless form as

$$r^* = \alpha r \approx \frac{\log_e R_0}{1 + (\alpha M)^{-1}} \qquad (6.11)$$

First, species with similar R_{0m} values should have similar θ^* values, independent of body size. Suppose that a collection of species of differing body size have αM independent of population size and have

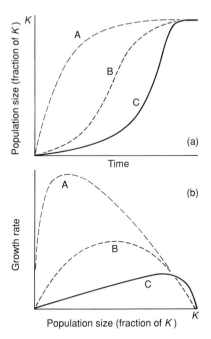

Fig. 6.5 (a) Population growth trajectories and (b) corresponding population growth rate curves showing variability in the position of the inflection points. As fractions of K these points are $0.2K$, $0.5K$, and $0.8K$ for curves A, B, and C, respectively. Although there is a continuum in the position of the inflection point, A approximates the dynamics of the productive commercial fish species and insect pests and C that of some large mammals. The intermediate case B represents a variety of cases including some protozoans. Redrawn from Fowler (1988).

the *same* curve for the decline of $\log_e R_0$ with relative population size θ. As illustrated in Fig. 6.7, $\log_e R_0$ is assumed to go from $\log_e R_{0m}$ at $\theta = 0$ to zero at $\theta = 1$. Equation (6.11) states that, for fixed αM, r^* is a unique function of $\log_e R_0$. As $\log_e R_0$ declines with θ, r^* also goes from its maximum value to zero along a unique path for each $\log_e R_0$ versus θ curve; thus all species in our collection will share the same r^* versus θ curve.

Now suppose that we wish to find θ^*, the θ associated with the largest dN/dT (for a particular species). We write

$$\frac{dN}{dT} = f(\theta) \quad (= rN) \ .$$

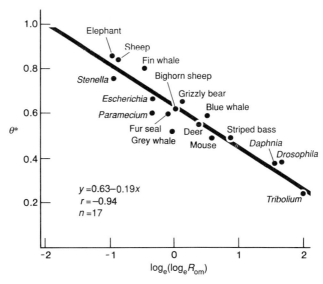

Fig. 6.6 The relationship between the relative position of the inflection point $\theta^* = $ (fraction of equilibrium levels) and R_{0m} for various species of animals as based on literature data. After Fowler 1988; data summarized in Table 6.1.

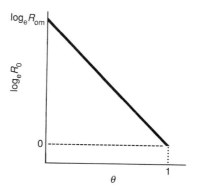

Fig. 6.7 The net reproductive rate R_0 declines with relative population size $\theta\ (= N/K)$. As N goes to the asymptotic size K (or $\theta \rightarrow 1$), $\log_e R_0 \rightarrow 0$. As developed in the text, this curve plays a key role in Fowler's first rule for population dynamics (species with the same R_{0m} have the same θ^*, where θ^* is the θ value which maximizes dN/dT).

Clearly θ^* is where $df/d\theta = 0$. However, the θ^* value is unchanged if we multiply $f(\theta)\ (= dN/dT)$ by any positive number. Assume that α is not altered by population density and write $f(\theta) = Nr(\theta) \propto \theta r^*(\theta)$.

θ^* maximizes $\theta r^*(\theta)$, but all the species in our collection share the same $r^*(\theta)$ function and thus they will have the same θ^* value. The basis of Fowler's first rule is that species with the same R_{0m} have the same decline in $\log_e R_0$ as θ goes from zero to unity. This is quite plausible, as in all cases $\log_e R_0$ must go from $\log_e R_{0m}$ to zero as θ goes from zero to unity. This result for θ^* is also independent of the value of αM.

Fowler's second rule is the specific negative relation between θ^* and R_{0m}, as illustrated in Fig. 6.6. This relation is for species occupying different R_0 versus θ curves with different R_{0m} values. These curves must also differ in *shape*. Suppose that $\log_e R_0$ has the general form $\log_e R_0 = (\log_e R_{0m})(1 - \theta)^u$. Since θ^* maximizes $\theta r^*(\theta)$, we can claim by eqn. (6.11) (with αM constant) that it maximizes $\theta(1 - \theta)^u (\log_e R_{0m})$. This makes $\theta^* = 1/(1 + u)$. By Fowler's fitted equation we have $\theta^* = 0.633 - 0.187 (\log_e (\log_e R_{0m}))$. Setting this function for θ^* equal to $1/(1 + u)$ and solving for the shape parameter u yields

$$u = \frac{0.367 + 0.187y}{0.633 - 0.187y}$$

where $y = \log_e (\log_e R_{0m})$. This means that species with $R_{0m} > 8$ are predicted to show *concave* $\log_e R_0$ versus θ curves ($u > 1$), while species with $R_{0m} < 8$ should show *convex* curves ($u < 1$), independent of body size. This hypothesis about shape is yet to be tested directly, as Fowler's data were used to generate it.

6.4 Summary

The r_{max} allometry results discussed here depend on the existence of three symmetries: first, R_{0m} is invariant with body size; second and third, α and M are invariant with population size but show ± 0.25 scalings with body size. Violations of one or more of these symmetries (α, M, R_{0m}) may lead to interesting predictions as to why species deviate from the typical r_{max} scaling lines. It may also prove interesting to ask how much these symmetry assumptions can be violated and still produce at least approximate -0.25 r_{max} allometry.

The predicted height of the r_{max} allometry depends strongly upon the height A of the production–growth function $dW/dT = A W^{0.75}$. Ectotherms show much slower growth (A small) contrasted with endotherms (A large); the ectotherms are thus predicted to have a lower r_{max} allometry (unless compensated for by very large R_{0m} values). r_{max}

for primates should also be smaller than r_{max} for other mammals of the same body size.

Fowler's rules for population dynamics suggest another kind of invariant: species with the same R_{0m} will show the same *shape* for the decline in R_0 with relative population size θ ($= N/K$). The shape itself will be related to R_{0m}: for $R_{0m} > 8$ the curves will be concave, while $R_{0m} < 8$ will lead to convex $\log_e R_0$ versus θ curves.

R_{0m} (and R_0) is a parameter that has found much use in epidemiology, where it is defined to be the number of secondary infections generated per primary infection (Anderson and May 1991); clearly $R_{0m} > 1$ is necessary for a disease to spread. Anderson and May discuss many examples of the usefulness of the R_0 parameter for understanding the dynamics of human disease. The theory developed in this chapter suggests that R_0 (and R_{0m}) is likewise useful for understanding other problems in the dynamics of populations.

7

Senescence (ageing)

7.1 Introduction

The Medawar–Williams evolutionary theory of senescence (or ageing) (Medawar 1952; Williams 1957) is based on the fundamental observation that an individual's future reproduction is worth less than its present reproduction simply because the individual is less likely to be alive to receive the future benefits (reviewed by Kirkwood and Rose (1991), Rose (1991), and Stearns (1992)). Of course, this is an asymmetry with respect to time. A useful definition of ageing is 'a persistent decline in the age specific fitness components of an organism due to internal physiological deterioration' (Rose 1991). The force of natural selection declines with age (formal arguments are given by Hamilton (1966), Emlen (1970), and Charlesworth (1980); the reader can convince himself or herself of this by looking at $\partial R_0/\partial b_x$ or $\partial R_0/\partial M_x$ for various and increasing ages x), and so such deterioration may generally be due to (1) the accumulation of detrimental mutations acting late in life (i.e. very weak selection against them) and/or (2) antagonistic pleitropy (a trade-off) where genes acting to increase fitness early in life cause inescapable detriments to function later in life. Rose (1991) discusses evidence bearing on each of these and proposes systems/experiments to test various elements. He also discusses various physiological and biochemical measures which have been developed (proposed) to measure ageing; we shall not discuss these measures here but ask the less technical question of what kind of organisms we should make measurements on.

An important aspect of Williams's (1957) contribution was to point out various within- and between-species comparisons that may be useful to test the theory (Rose 1991, p. 91). This chapter points towards four biological systems which may prove useful for testing two of Williams's ideas. The first idea is that 'aging should be more rapid in those organisms that do not increase markedly in fecundity after maturity than those that do show such an increase'. The second idea is that senescence should not begin until after the age at maturity.

The four systems discussed here are (1) determinate versus indeterminate growers, (2) sex-changing fish, (3) pollen grains, and (4) fish with alternative male life histories.

7.2 Determinate versus indeterminate growth

Figure 7.1 is a very simplified view of a survivorship $L(x)$ curve in the face of senescence; the adult instantaneous mortality rate M is approximately constant until some maximum age T_{max}. There are two time-scales in this simple world: $1/M$ and $T_{max}^* = T_{max} - \alpha$. We use T_{max}^* instead of simply T_{max} since T_{max}^* measures the maximum time left after maturation (at age α) and is the maximum adult life-span. The average adult life-span will be approximately $1/M$ provided T_{max}^* is greater than about $2/M$, i.e. the average adult life-span will not be strongly affected by the maximum life-span T_{max}^*. It is interesting to ask whether $T_{max}^*/(1/M)$ ($= MT_{max}^*$) takes on characteristic values for different kinds of organisms. In particular, MT_{max}^* might be expected

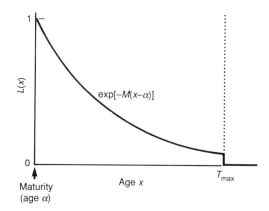

Fig. 7.1 A simple $L(x)$ schedule with senescence. Mortality M is constant after maturity (age α) until age T_{max} when any individuals still alive die of 'old age'. The text develops the theme that various kinds of organisms may show characteristic ratios for the average ($\approx 1/M$) versus the maximum ($T_{max} - \alpha = T_{max}^*$) adult life-span, where the dimensionless number is defined as $T_{max}^*/(1/M)$ or MT_{max}^*; in particular, it is argued that MT_{max}^* ought to be larger where fecundity increases substantially after maturation, for example with indeterminate (versus determinate) growth.

to be smaller for determinate versus indeterminate growers, since indeterminate growers usually increase markedly in fecundity after maturity. This hypothesis is based on the assumption that selection increases the fecundity schedule at a cost of lowered T^*_{max} and that M is externally imposed on the organism. MT^*_{max} can be considered a dimensionless measure of ageing since it relates the maximum to the average adult life-span. One comparison, that of fish versus mammals, is easily made using data discussed in Chapters 4 and 5. It should be noted that $MT^*_{max} = (T_{max} - \alpha)M = MT_{max} - \alpha M$.

Beverton (1963) estimated M and T_{max} for field populations of a wide variety of fish. To summarize his extensive analysis, (and that of Hoenig (1983)) $T_{max} \approx 5/M$–$6/M$ and thus $MT_{max} \approx 5$–6. Since $\alpha M \approx 2$ for fish, we have $MT^*_{max} \approx 3$–4.

What about mammals? As mentioned in the caption to Fig. 5.6, we can use field life-table data to estimate M and combine this with zoo data for T_{max} (Eisenberg 1981) to yield the relation (units of years) $T_{max} = 2.5/M + 0.25$. Since $\alpha M \approx 0.75$ for mammals, we have $MT^*_{max} \approx 1.75$ for mammals.

The indeterminate growers (fish) have the larger MT^*_{max} numbers (3–4 versus 1.75). The analysis is admittedly rather rough, but hopefully it will serve as an invitation to more precise work in the future.

7.3 Sex-changing fish

Many fish and invertebrates have fecundity almost proportional to weight and weight increasing as $\{1 - \exp[-k(\text{age})]\}^3$. While this makes the future worth a great deal (as just discussed), there is one kind of organism which may hold a record for the potential value of reproduction (fecundity) at the very end of its life; this is a protogynous female to male sex-changer with a breeding sex ratio of, say, ten females to one male (the extremes in Fig. 2.7). In order to see how a sex-changer with such a high sex ratio (10:1) ought to be useful to test ageing theory, consider Fig. 7.2, which is a standard $L(x)$ curve (with age zero at maturity). Let τ be the age of sex change. The stippled area marked ① is the area under the curve before τ, i.e. the time spent as a female. The area marked ② is the area under the curve after τ, i.e. the time spent as a male. It is easy to show (e.g. Chapter 2) that in evolutionary equilibrium with a non-growing population and a stable age distribution, the adult sex ratio ♀/♂ is the ratio of the areas, or

$$\text{♀}/\text{♂} = \text{①}/\text{②} \; .$$

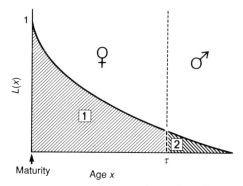

Fig. 7.2 Some protogynous sex-changers have breeding sex ratios ♀/♂ of about 10:1; this is the ratio of the female to male areas under the $L(x)$ curve (sex change at age τ). Since reproductive success is equal for the two sexes, fecundity is very high during the male phase at the end of the life-span. This life history should show comparatively little ageing in contrast with a dioecious species with similar average mortality rates.

Suppose that we have ♀/♂ $= 10$ and an adult instantaneous mortality rate of M. Then

$$\frac{\displaystyle\int_0^{\tau} e^{-Mx}\, dx}{\displaystyle\int_{\tau}^{\infty} e^{-Mx}\, dx} = 10$$

which gives $M\tau = 2.4$. $M\tau$ is a dimensionless number which thus takes on a fixed (invariant) value for species with the same adult sex ratio.

The above analysis implies that about 9 per cent of the adults live to become males ($e^{-M\tau} = e^{-2.4} = 0.09$). The average adult life-span is $1/M$ years; thus τ is at 2.4 times the average adult life-span, a rather old age. The importance of this becomes clear when we realize that in evolutionary equilibrium half of an individual's reproduction (fecundity) happens after age τ, a simple consequence of each zygote receiving half its genes from males and half from females, i.e. the Fisher inheritance symmetry of Chapter 2. With a breeding sex ratio of 10:1, a great deal of fecundity awaits the second sex period, very late in life. Let us contrast this with a fish that does not change sex but has the same demography. Suppose that females of this species have an average fecundity of b_1 before age τ and b_2 after. How large must b_2 be to match the comparable sex-changer? For simplicity let b_1 and b_2 be constants. From maturity to τ this female will produce

$b_1 \int_0^\tau e^{-Mx} \, dx$ eggs; after τ, she will produce $b_2 \int_\tau^\infty e^{-Mx} \, dx$ eggs. If

$M\tau = 2.4$ and if the ratio of egg production before τ to egg production after τ is equal to unity (equal numbers of eggs before and after τ), then $b_2 > 10b_1$. This comparable dioecious female must have an average fecundity at the end of her life about 10 times greater than that in her early life. Beverton and Holt (1959) note that cod (Gadidae) begin reproducing at about 17 per cent of their asymptotic weight. If fecundity is proportional to weight and we assign elderly cod the asymptote, the *largest possible* b_2/b_1 is $1/0.17 = 6$, only about half the value of 10 assigned to the sex changer. Cod are unusual in their amount of post-maturity growth. A more typical dioecious fish starts at 25–30 per cent of the asymptotic weight; b_2/b_1 is thus more usually less than 3–4, nowhere near the 10 for the sex changer.

Sex-changing fish with extreme breeding sex ratios should be very useful subjects in ageing studies. An individual of such a species has a great deal of fecundity in a short period at the very end of its life. They should show much less senescence than dioecious species with similar average mortality rates.

7.4 Pollen grains

If protogynous fish represent greatly increased fecundity with age, pollen represents the opposite. A pollen grain awaiting pick-up or ready to be released to wind or water is faced with a dichotomous future because in the next short time period (minutes, hours, days at most) it will either (1) end up on a stigma and be competing to fertilize the locally available ovules or (2) be transported to somewhere other than a conspecific stigma and will thus be dead, having virtually no chance of being repositioned to an appropriate stigma. The original transport is probably mostly related to time, with time or age zero when the pollen grain is physiologically an individual. Pollen is quite difficult to keep alive for more than a few days, suggesting that its design represents quite clearly this time-dependent dichotomous event of 'on the stigma, or bust'. It seems probable that pollen life-times are designed as an approximate match to the time course of the time to transport event, which may differ depending upon the type of vector (or other transport means). We must also recognize that pollen design represents selection on the mother to divide resources optimally among her many pollen grains, and that pollen design may reflect to some degree selection for efficient transport itself.

Why does pollen die? There are two possibilities. Pollen may become non-viable due to exhaustion of food (or water) stores or the breakdown of what can be termed competitive biochemical readiness. A pollen grain is a male in a high state of competitive readiness, designed to grow towards ovules after landing upon the stigma. It seems quite possible that the trade-off for a high level of readiness is that the state cannot be maintained for a long time; thus an age-dependent trade-off would exist and natural selection would push the readiness up at the cost of a shorter effective life-time. Just as a pollen grain might be given only enough food to live past the transport time, its growth machine ought to decay on a similar time-scale (particularly if the age versus readiness trade-off is real.) Even within various transport means, useful ageing comparisons seem possible. Cox (1988 and personal communication) has suggested that the transport time for hydrophilious (water-borne) pollination should differ dramatically between freshwater and intertidal-living angiosperms, since tidal cycles effectively limit the time available for pollen usefulness. Freshwater lakes have no tidal cycle to flush pollen away, and so we expect much longer pollen viability in fresh water.

7.5 Alternative male life histories

Rose (1991) advocates laboratory selection experiments to produce within-species lines with greatly altered life-spans; his own work has focused on analysis of such strains for *Drosophila*. In this section we suggest the use of natural variation in life-span caused by differing environmental regimes. One obvious candidate is latitudinal variation for fish and marine invertebrates. For example, the shrimp *Pandalus borealis* has a maximum life-span of 3–4 years at a latitude of about 45°N, but 7–8 years in the Bering Sea or off Spitsbergen (Charnov 1979c). On land, the rattlesnake *Crotalus viridis* has an age of first breeding of 3 years in Utah, but 7 years in British Columbia (annual adult mortality rates are 25 per cent and 15 per cent respectively) (Shine and Charnov 1992). Many more such within-species between-location comparisons are undoubtedly possible; however, an even more powerful comparison would be between alternative life histories within a single population of one species.

Males of the common North American bluegill sunfish (*Lepomis macrochirus*) certainly fit this bill (Chapter 3; Gross and Charnov 1980; Gross 1991a). Frequency-dependent sexual selection (mating advantage) has produced two distinct non-overlapping (in age) male life histories as shown in Fig. 3.1. In Lake Opinicon, near Kingston, Ontario, Canada, about 20 per cent of a bluegill male cohort begin reproducing at age 2 years, acting as cuckolders at the nests of parental males; these small males do not live beyond age 6 years. The parental males do not begin reproducing until age 7–8 years and do not survive beyond age 11 years. What makes this system of much interest is that, in evolutionary equilibrium, the life-time fitness is equal for individuals in each male pathway; data in support of this identity were discussed in Chapter 3. Physiological comparisons between the morphs may prove useful in ageing studies, particularly to test the idea that ageing does not begin until maturity; the pathways vary greatly in the age at maturity (2 versus 8 years), yet each yields the same life-time fitness. Each pathway also yields about the same average adult life-span (*c.* 3–4 years).

7.6 Summary

The decline in the force of natural selection with age is one of the great asymmetries in population biology. It is the general cause of what we label senescence or ageing (Rose 1991). In this chapter some implications of this have been discussed and biological systems that seem particular useful for ageing studies have been indicated.

The MT^*_{max} number is a fairly rough measure for ageing; more detailed descriptions of $L(x)$ curves should yield more precise measures. However, preliminary calculations suggest that MT^*_{max} is larger for indeterminate versus determinate growers, which is expected if ageing is postponed when fecundity increases greatly with adult age. Sex-changing fish with extreme adult sex ratios also have a great deal of fecundity available at old adult ages. In contrast, pollen grains have no reproductive opportunities after they have been transported, if they end up on any surface other than an appropriate stigma. Some fish (e.g. bluegill sunfish) have alternative male life histories, where the two pathways differ greatly in age but in evolutionary equilibrium yield the same fitness to an individual. Ageing should be much slower on the pathway where reproduction is postponed to older ages.

8

Finis

This final chapter will not repeat the discussions from the previous seven. I hope that I have convinced the reader of the utility of a *symmetry approach to life histories* even if there is disagreement with particular modelling assumptions (e.g. the yearly clutch size model for mammals is probably too simple). Invariance under particular transformations is a worthy consideration; the data support the reality of invariance. Open questions and future directions include the following.

1. The stationary population assumption, with the immense usefulness of the stable age distribution as well as R_0 as a fitness measure, seems almost too good to be true. Let us just say it works and wonder why. Perhaps it is merely a useful first-order approximation to a stochastic fitness measure, which explicitly considers the fluctuating populations that we all know really characterize nature (Metz *et al.* 1992). Or perhaps the population fluctuations are mere noise and R_0 is an approximately correct fitness measure when averaged over several generations ('the medium run') as well as being correct for the population dynamic assumption of $R_0 \approx 1$. Perhaps the assumption of a stable age distribution is likewise reasonable as an average in the medium run. Theorists could help here by tackling the question of when R_0 is a poor fitness choice. It does not help to point out that stationarity is commonly violated in nature. We all know this, but what we do not know is when the violations really matter with respect to the life history problems, particularly R_0 as a fitness measure. A good analogy is the usefulness of single-locus population genetics for obtaining ESS results. No one believes that the characters are commonly controlled by a single locus (they are usually polygenic). What we do believe is that the ESS answers are approximately independent of the details of the genetic system and we are simply using the single-locus argument to find the equilibrium phenotype (Maynard Smith 1982).

2. Consider the invariance of α, b, M (and individual production) with changes in population size, putting all density dependence into survival of the very young. This is a modelling assumption (or the outcome of natural selection on α in the face of mortality and density independence of individual production rates) and a hypothesis to be tested. Data in support of it exist for mammals (Fowler 1981, 1987; Sinclair 1989), and the assumption underlies almost all fishery theory where adult mortality and individual growth are typically assumed to be density-independent relations. How generally true it is, or whether other general forms of population size invariance exist, remains to be shown. Perhaps the rules for density dependence will commonly be in the form of *where in the life history it is present*, rather than what specific biological agents cause it.

3. A major tool of this book is to characterize life histories by using dimensionless numbers to display relationships between elements of the life history; this allows us to see squirrels and elephants as the same (e.g. the same αM and δ). Dimensionless variables are widely used in dynamic problems (e.g. relative fitness in population genetics models) and provide 'criteria of similarity' for objects of different size etc. (Stahl 1962). Stephens and Dunbar (1993) review formal dimensional analysis and its usefulness in behavioural (evolutionary) ecology. Consider two examples from population dynamics with determinate growth. The net reproductive rate R_0 can be written as (from eqn. (6.2))

$$R_0 = \frac{bS(\alpha)}{M}$$

or, in alternative dimensionless form,

$$R_0 = \frac{\alpha b}{\alpha M} e^{-\bar{Z}\alpha} \ . \tag{8.1}$$

Equation (8.1) shows one relationship between the dimensionless numbers R_0, αM, αb, and $\alpha \bar{Z}$.

Equation (6.4) gives us the intrinsic rate of increase r for a determinate grower; αr is a dimensionless intrinsic rate of increase. Equation (6.4) can be rewritten as

$$e^{\alpha r}(\alpha M + \alpha r) - \alpha b e^{-\bar{Z}\alpha} = 0 \ . \tag{8.2}$$

Equation (8.2) gives a relationship between the four dimensionless numbers αr, αb, αM, and $\alpha \overline{Z}$. If, as developed in this book, only $\overline{Z}$ changes with population size, then these equations lead to some general rules for population dynamics (as shown in Chapter 6). Equations (8.1) and (8.2) also suggest that the natural time-scale for a life history is units of α; for example, αr is the intrinsic rate of increase in one unit of this time. Species with different values of α may show very similar population dynamics when each is viewed in its own time frame (i.e. we compare values of αr between species).

4. The life history modelling of Chapters 1 and 4–6 assumed constant (age-independent) adult mortality. Tests of hypotheses used the average adult instantaneous mortality rate (Beverton 1963), which was calculated on the stable age distribution. This average mortality rate M is almost identical with the inverse of the expectation of further life at the age at maturity (treating the $L(x)$ schedule as a continuous curve). This is why $1/M$ has been termed the average adult life-span. The same definition of M is used for all taxa so that between-taxa comparisons for numbers such as αM would not turn up differences simply because of the use of different statistics for M. Life history evolution models need not assume this type II mortality (M a constant), and it would be interesting to see how robust the predicted results are to other life-table assumptions.

5. The Fisherian inheritance symmetry (one mother and one father) underlies all sex allocation evolution (Charnov 1982; Bull and Charnov 1988). We either know that it is present, or else we know the way that it is violated. It remains to be seen if any equally strong symmetries exist for other life history problems. In this book we elect to act as if the answer is in the affirmative and to search for them. We should never forget that the Fisher symmetry is true, and is perhaps the best example (after Mendelism) of the usefulness of symmetry principles in evolutionary ecology.

6. In the spirit of invariance, one of the deepest puzzles is the source of the shape symmetries for trade-offs, as developed in Chapters 1, 2, 4, 5, and 6. While the choice of power functions for trade-offs is perhaps too simple, these functions have the interesting attribute that the exponents (the shape parameters) are dimensionless. Thus if we solve a life history evolution problem in dimensionless form (e.g. αM), it almost follows automatically that only the exponents will appear in the final equations (αM = something about shape). All units are carried by the intercept (height) parameters, and so for them to

appear in a dimensionless equation would require other parameters to cancel the units. While nothing precludes this and examples of it can be constructed, it is common for only shapes to remain in the ESS answers. But what sets the shapes, and causes their invariance? It may be a useful exercise to non-dimensionalize other life history evolution theories and ask what is implied about trade-offs if the dimensionless numbers of these theories are invariants.

7. The dimensionless number αM appeared in several chapters. Suppose that the complications of indeterminate growth are ignored and even fish, shrimp, and reptiles are treated as if growth ceased upon attainment of maturity. Then if growth follows the 0.75 scaling of production with body size (but see below), the mammal argument of Chapter 5 may apply even for the fish etc., particularly if mortality stabilizes prior to age α. This suggests that we apply the ESS results for αM (eqn. 5.9(b)) to these other groups. The ESS is $\alpha M = 3(1 - \delta^{0.25})$; Fig. 8.1 shows the theoretical αM versus δ. The extreme sensitivity of αM to δ near zero should be noted; indeed αM drops from 3 to 2 as δ goes from zero to 0.01. Figure 8.2 linearizes the αM relation by plotting it versus $\delta^{0.25}$. We already know (Fig. 5.7 and text discussion) that the mammals cluster around the theoretical line, with the observed average quite close. $\delta^{0.25}$ has also been estimated for

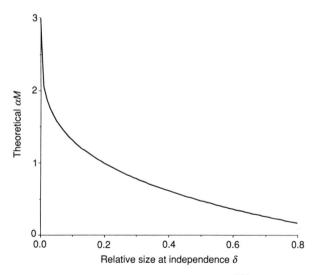

Fig. 8.1 Theoretical αM versus δ (or $\alpha M = 3(1 - \delta^{0.25})$). It should be noted that αM is very sensitive to δ near zero; indeed αM is 3 at $\delta = 0$ but drops to 2 at $\delta = 0.01$.

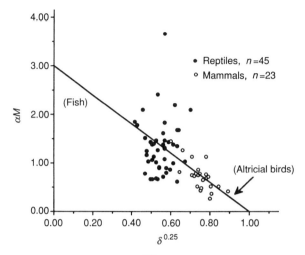

Fig. 8.2 Theoretical αM versus $\delta^{0.25}$, treating all species as if they have determinate growth. The various groups (birds, mammals, reptiles, fish) do indeed have approximately the theoretically predicted αM for their respective δ values. See text discussion for qualifications on this conclusion.

the 45 reptile (lizard and snake) populations. This was done by noting that $\delta = W_0/W_\alpha$ should equal $(\ell_0/\ell_\alpha)^3$; thus the reptiles of Chapter 4 yield $\delta^{0.25} = (\ell_0/\ell_\alpha)^{0.75}$. As shown in the figure, the reptiles do not vary greatly in $\delta^{0.25}$ (their $\delta^{0.25}$ range is about half the size of the mammal range). While there is quite a lot of variation in αM, the average $\delta^{0.25}$ for these reptiles is 0.55 and predicts quite closely the average αM, estimated in Fig. 4.12 to be 1.32 (or $3(1-0.55) = 1.35$). $\delta^{0.25}$ must be much smaller for turtles, and so an interesting comparison seems possible.

The fish (and shrimp) have $\delta \to 0$ and $\alpha M \approx 2$, or above; is 2 too small, compared with 3, for the fish to be claimed to fit the relation? αM is so very sensitive to δ near zero that this is hard to answer. What we can claim is that the fish (and shrimp) have the smallest values of δ and αM near the prediction of the 0.75 production theory. However, not all fish have small δ, and a comparison of the αM values for live-bearers with large offspring (e.g. sharks) might help settle the issue: αM is in the bird range for live-bearers of the genus *Sebastes* (Beverton 1992); $\alpha M \approx 2$ with $\delta \approx 0$ is predicted with determinate growth if the exponent of the production relation is 2/3 instead of 3/4; eqn. (5.14) gives an ESS αM of $2(1 - \delta^{0.33})$. Perhaps aquatic ectotherms have a 0.67 production scaling relation. However, this 0.67 scaling cannot apply to the reptiles since it leads to an incorrect αM value (*c.* 1.1 versus the observed value of 1.32).

Finally we have the birds, particularly altricial birds which contributed almost all the data in Fig. 1.8. They have $\alpha M \approx 0.4$ on average, the smallest αM of all. They also have the largest relative size for their offspring at independence; indeed some species have fledglings of near adult size ($\delta \to 1$!). Perhaps in some species the age at maturity is at the very next good season; recall that seasonality was not considered in the theory so that $\alpha \to 0$ would essentially mean in practice the earliest possible. However, some birds present a much greater puzzle, one which questions their placement at all on this 'body size versus production' graph; they are near adult size at independence but none the less delay maturity for several years (discussion in Stearns (1992)). This is puzzling since it is not clear what they are waiting for (but guesses include acquiring experience in foraging or becoming able to compete for a territory or breeding place). This pattern also suggests that adult body size is adjusted to an optimum for adult function; the mammal model of Chapter 5 had no optimum adult body size independent of the time and mortality cost to reach it. Therefore birds have been placed at a large δ on Fig. 8.2, but the question remains: is the assignment reasonable or even correct? If not, then why is αM so small for them? Just *what* increases slowly with a delay in maturation, as developed in Chapter 1?

Setting caution aside, the fit between αM and δ is rather encouraging; at the least it suggests that we are going in the right direction through the use of production-based models to generate life histories. Figure 8.2 also includes both ectotherms and endotherms, which have quite different production versus body size relations (Chapter 6) but the αM relation seems to obey the same δ-based rule. Again, this might be expected, since a dimensionless prediction like αM ought to be invariant to the actual heights of the production relations. However, it is unclear just why indeterminate growth does not spoil the result which was derived with determinate growth, but then we have a good puzzle for future work. It is also not clear how to connect this αM result to the approach developed in Chapter 4; perhaps these puzzles will be answered when someone shows how to turn a production relation like $dW/dT = AW^{0.75}$ into a growth equation like the Bertalanffy equation through natural selection on α and the ever-increasing relative allocation of resources to reproduction after α. A step in this direction is given next.

8. Let us ignore for a moment the evolution of the indeterminate growth curve and take a more mechanistic view. The Bertalanffy curve for weight near zero behaves like the production relation $dW/dT = AW^{0.67}$. Suppose that this growth curve applies until the age at maturity

α after which growth slows down only because of the allocation of some resources to offspring production (Roff 1982, 1983). A common pattern in fish data is for egg production to be proportional to body weight (Roff 1982, 1983). If we assume this true, then production per unit time EW is given to reproduction after age α. Thus growth after age α will follow the rule

$$\mathrm{d}W/\mathrm{d}T = AW^{0.67} - EW \ . \tag{8.3}$$

Of course, this equation is the differential form of Bertalanffy's equation (e.g. Reiss 1989). Growth prior to α will follow $\mathrm{d}W/\mathrm{d}T = AW^{0.67}$ and will only approximate Bertalanffy growth. Growth in length will have a constant slope before age α.

This growth model is illustrated in Fig. 8.3. Growth ceases when $AW^{0.67} - EW = 0$; the associated W is the asymptotic weight W_∞. Weight W_α at age α is found by solving $\mathrm{d}W/\mathrm{d}T = AW^{0.67}$; if weight is zero at time zero, this yields $W_\alpha^{0.33} = 0.33A\alpha$.

What have we gained by using this approach to indeterminate growth? One interesting dimensionless prediction follows immediately from the calculations in the last paragraph. Assume that length follows weight to the power 0.33. Then

$$\left(\frac{W_\alpha}{W_\infty}\right)^{0.33} = \frac{\ell_\alpha}{\ell_\infty} = 0.33\,E\alpha \ . \tag{8.4}$$

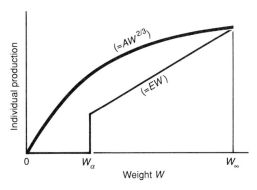

Fig. 8.3 A model for indeterminate growth. Prior to age α growth follows the production relation $AW^{0.67}$. After α, EW of production is given to offspring so that growth now follows $AW^{0.67} - EW$. This growth hypothesis predicts that species with the same relative size at maturity W_α/W_∞ will share the same value for the dimensionless number $E\alpha$. The reproductive effort E will be inversely proportional to the age at maturity α. See text for detailed derivation.

Thus, a group of species having the same ℓ_α/ℓ_∞ value will also have the same $E\alpha$ value: E will be inversely proportional to the age at maturity α with the constant of proportionality equal to $3\ell_\alpha/\ell_\infty$. Our growth model thus yields a new invariant $E\alpha$ for species with an invariant ℓ_α/ℓ_∞ value. Of course, if such species also have invariant αM, they will likewise have constant E/M. As before, the height A of the basic production scaling does not appear in these new dimensionless predictions. Since various taxa of fish have different ℓ_α/ℓ_∞ values (Beverton 1992), tests of eqn. (8.4) seem quite possible. If true, then just why natural selection sets $E \propto M$ (or α^{-1}) becomes an interesting question; symmetry at the level of a trade-off is a good first guess. A growth model which defines a new invariant αE now requires the life history evolution model to output something else.

9. Does external mortality, combined with body size production relations, drive life histories (see also Stearns 1992, p. 208)? Many life history variables, particularly allometric residuals, correlate with mortality rates (Harvey and Pagel 1991), whereas few life history variables (corrected for body size) correlate with other features of ecology. Given a production relation $dW/dT = AW^{0.75}$ with A fixed, many different life histories will result from different external mortality sources, as shown in Chapter 5. Also, a great many more life histories can be generated if A is allowed to differ between taxa. The discussion in this book hints that much life history variation can be understood in terms of just a few numbers (A, δ, M) and the presence of determinate versus indeterminate growth. A mortality cost to reproduction is not on this list. Perhaps it should be.

If we assume a 0.75 production scaling, and approximate *all* life histories with the determinate growth model of Chapter 5, the following results (eqn. 5.7):

$$\text{Average adult lifespan} = \frac{1}{M} = \frac{1.33}{A} \cdot W(\alpha)^{0.25} \ .$$

At a fixed maturation body size ($W(\alpha)$), the average adult life-span (M^{-1}) is inversely proportional to the height (A) of the production function. Since, as argued in Chapter 6, ectotherms have much the smaller A values (see also Case 1978; Stearns 1992, his figure 5.2), they are expected to show the larger life-spans at any fixed body size. This is true in a comparison of lizards and snakes versus mammals; the reptile M^{-1} data from Fig. 4.12 may be associated with $W(\alpha)$ to show that a one kilogram reptile has M^{-1} 3-4 times larger than M^{-1} for a 1 kg (non-primate) mammal.

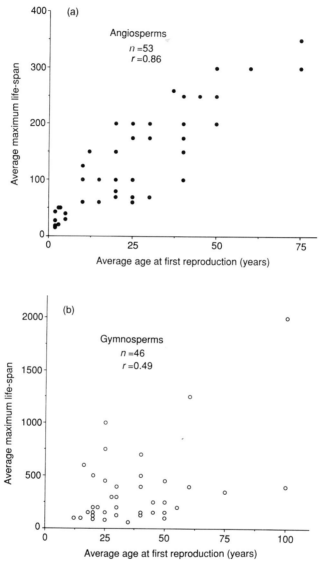

Fig. 8.4 Maximum life-span versus age at maturity within two large plant groups. Data from Loehle (1988).

10. Finally, what about plants? Loehle (1988) has compiled data on α and 'average maximum life-span' for 46 gymnosperm and 53 angiosperm (trees and shrubs) species. The data are plotted in Fig. 8.4; angiosperms

show an almost proportional relation whereas gymnosperms show almost no relation at all. Gymnosperms have most of the high y values; if the y variate is estimated in the same way for the two groups (Loehle is not really clear on how the maximum life-span was obtained), then gymnosperms have the greater y/x numbers. Why?

In Chapter 1 I suggested that it would often be easier to find invariants than to *explain* them. For example, I still do not know why b/M is an invariant for birds, as shown in Fig. 1.2 (Ricklefs 1969, 1977). The challenge is to use the invariance to probe the life history for deeper causal factors. Prior to this approach, it simply never would have occurred to me to derive a result such as the invariance of αE (this chapter) for species with similar ℓ_α/ℓ_∞ values. In this way, the αE number may help us to understand the structure of growth curves even if my particular model fails when confronted with data.

Glossary of major parameters and functions

The following parameters and functions are used several times in the book, often to define dimensionless numbers.

α age at maturity (first reproduction)

A height of the growth/production function ($dW/dT = AW^c$; $c \approx 0.75$)

b yearly clutch size, usually daughters per year

C efficiency parameter for female mammals investing in offspring (p. 94)

δ relative size at independence from mother; equals weight at weaning or hatching (W_0) divided by mother's weight at age α ($W(\alpha)$)

E reproductive effort (p. 142)

h slope of the assumed trade-off between k and ℓ_∞ (p. 77)

k growth coefficient in the Bertalanffy growth equation (Fig. 4.1)

$\ell\alpha$ length of an indeterminate grower at age α (Fig. 4.1)

ℓ_0 length at age zero (hatching) in the Bertalanffy growth equation (Fig. 4.9)

ℓ_∞ asymptotic length in the Bertalanffy growth equation (Fig. 4.1)

$L(x)$ proportion of individuals surviving to age x

M instantaneous mortality rate of adults (survival for one year $= e^{-M(\text{one year})}$)

r intrinsic rate of population increase (p. 114)

R_0 net reproductive rate (defined on p. 8)

R_{0m} R_0 maximum, shown in rarified populations (p. 119)

$S(\alpha)$ proportion of offspring surviving to age α

τ age at sex (or sex ratio) change

$T_{\max}$ oldest observed age

$V(\alpha)$ reproductive value at age α (p. 8)

W body weight

$W(\alpha)$ body weight at age α

W_0 body weight at independence from mother (weaning, hatching)

$Z(x)$ instantaneous mortality rate at age x for immatures (i.e. $x < \alpha$)

$\overline{Z}$ average of Z, defined so that $e^{-\overline{Z}\alpha} = S(\alpha)$ (p. 97)

References

Adams, J., Greenwood, P., and Naylor, C., 1987. Evolutionary aspects of environmental sex determination. *Int. J. Invertebr. Reprod.* **11**: 123-36.

Alexander, R. D., Hoogland, J. L., Howard, R. D., Noonan, K. M., and Sherman, P. W., 1979. Sexual dimorphism and breeding system in pinnipeds, ungulates, primates and humans. In Chagnon, N. and Irons, W. (eds.), *Evolutionary biology and human social behavior*, pp. 402-35. Wadsworth.

Anderson, R. M. and May, R. M., 1991. *Infectious diseases of humans*. Oxford University Press.

Andrews, R. M., 1982. Patterns of growth in reptiles. In Gans, C. and Pough, F. H. (eds.), *Biology of the reptilia*, Vol. 13, pp. 273-320. Academic Press, New York.

Arak, A., 1988. Sexual dimorphism in body size: a model and a test. *Evolution* **42**: 820-5.

Banse, K. and Mosher, S., 1980. Adult body mass and annual production/biomass relationships of field populations. *Ecol. Monogr.* **50**: 355-79.

Barnard, C. J. (ed.), 1984. *Producers and scroungers: strategies for exploitation and parasitism*. Croom Helm, London.

Bell, G., 1980. The costs of reproduction and their consequences. *Am. Nat.* **116**: 45-76.

Bell, G. and Koufopanou, V., 1986. The cost of reproduction. *Oxford Surv. Evol. Biol.* **3**: 83-131.

Berrigan, D., Purvis, A., Harvey, P. H., and Charnov, E. L., 1993. Phylogenetic contrasts and the evolution of mammalian life histories. *Evol. Ecol.* **7**: 270-8.

Berry, J. F. and Shine, R., 1980. Sexual size dimorphism and sexual selection in turtles (order Chelonia). *Oecologia* **44**: 185-91.

Beverton, R. J. H., 1963. Maturation, growth and mortality of clupeid and engraulid stocks in relation to fishing. *Rapp. P.-V. Reun. Cons. Int. Explor. Mer.* **154**: 44-67.

Beverton, R. J. H., 1987. Longevity in fish: some ecological and evolutionary considerations. In Woodhead, A. D. and Thompson, K. H. (eds.), *Evolution of longevity in animals*, pp. 161-86. Plenum Press, New York.

Beverton, R. J. H., 1992. Patterns of reproductive strategy parameters in some marine teleost fishes. *J. Fish Biol.* **41** (Supplement): 137-60.

Beverton, R. J. H. and Holt, S. J., 1959. A review of the lifespans and mortality rates of fish in nature and the relation to growth and other physiological characteristics. In *Ciba Foundation colloquia in ageing. V. The lifespan of animals*, pp. 142-77. Churchill, London.

Bilton, H. T., 1980. Returns of adult coho salmon in relation to mean size and time at release of juveniles to the catch and escapement. *Can. Tech. Rep. Fisheries and Aquatic Sciences 941*.

Blackmore, M. and Charnov, E. L., 1989. Adaptive variation in environmental sex determination in a nematode. *Am. Nat.* **134**: 817–23.

Blueweiss, L., Fox, H., Kudzma, V., Nakashima, D., Peters, R., and Sams, S., 1978. Relationship between body size and some life history parameters. *Oecologia* **37**: 257–72.

Borghetti, J. R., Iwamoto, R. N., Hardy, R. W., and Sower, S., 1989. The effects of naturally occurring androgens in practical diets fed to normal-sired and jack-sired progeny of coho salmon (*Oncorhynchus kisutch*). *Aquaculture* **77**: 51–60.

Brockmann, H. J. and Grafen, A., 1992. Sex ratio and life-history patterns of a solitary wasp *Trypoxylon politum* (Hymenoptera: Sphecidae). *Behav. Ecol. Sociobiol.* **30**: 7–27.

Brody, S., 1964. *Bioenergetics and growth*. Hafner, New York.

Brunet, J., 1992. Sex allocation in hermaphroditic plants. *Trends Ecol. Evol.* **7**: 79–84.

Bull, J. J., 1980. Sex determination in reptiles. *Q. Rev. Biol.* **55**: 3–21.

Bull, J. J., 1981. Sex ratio evolution when fitness varies. *Heredity* **46**: 9–26.

Bull, J. J., 1983. *Evolution of sex determining mechanisms*. Benjamin/Cummings, Menlo Park, CA.

Bull, J. J. and Charnov, E. L., 1977. Changes in the heterogametic mechanism of sex determination. *Heredity* **39**: 1–14.

Bull, J. J. and Charnov, E. L., 1988. How fundamental are Fisherian sex ratios? *Oxford Surv. Evol. Biol.* **5**: 96–135.

Bull, J. J. and Charnov, E. L., 1989. Enigmatic reptilian sex ratios. *Evolution* **43**: 1561–6.

Butler, T. H., 1964. Growth, reproduction and distribution of Pandalid shrimps in British Columbia. *J. Fish. Res. Board Canada* **21**: 1403–52.

Calder, W. A., 1984. *Size, function and life history*. Harvard University Press.

Case, T. J., 1978. On the evolution and adaptive significance of postnatal growth rates in the terrestrial vertebrates. *Q. Rev. Biol.* **53**: 243–82.

Caughley, G. and Krebs, C. J., 1983. Are big mammals simply little mammals writ large? *Oecologia* **59**: 7–17.

Charlesworth, B., 1994. *Evolution in age structured populations* (2nd edn). Cambridge University Press.

Charlesworth, D. and Charlesworth, B., 1981. Allocation of resources to male and female functions in hermaphrodites. *Biol. J. Linn. Soc.* **15**: 57–74.

Charlesworth, D. and Charlesworth, B., 1987. The effect of investment in attractive structures on allocation to male and female functions in plants. *Evolution* **41**: 948–68.

Charlesworth, D. and Morgan, M., 1991. Allocation of resources to sex functions in flowering plants. *Phil. Trans. R. Soc. Lond. B* **332**: 91–102.

Charnov, E. L., 1979*a*. The genetical evolution of patterns of sexuality: Darwinian fitness. *Am. Nat.* **113**: 465–80.

Charnov, E. L., 1979*b*. Simultaneous hermaphroditism and sexual selection. *Proc. Natl. Acad. Sci. USA* **76**: 2480–4.

Charnov, E. L., 1979*c*. Natural selection and sex change in Pandalid shrimp: test of a life history theory. *Am. Nat.* **113**: 715–34.

Charnov, E. L., 1980. Sex allocation and local mate competition in barnacles. *Mar. Biol. Lett.* **1**: 269–72.

Charnov, E. L., 1982. *The theory of sex allocation*. Princeton University Press.

Charnov, E. L., 1986. Life history evolution in a 'recruitment population': why are adult mortality rates constant? *Oikos* **47**: 129–34.

Charnov, E. L., 1987. On sex allocation and selfing in higher plants. *Evol. Ecol.* **1**: 30–6.

Charnov, E. L., 1989*a*. Natural selection on age of maturity in shrimp. *Evol. Ecol.* **3**: 236–9.

Charnov, E. L 1989*b*. Evolution of the breeding sex ratio under partial sex change. *Evolution* **43**: 1559–61.

Charnov, E. L., 1990. On evolution of age of maturity and the adult lifespan. *J. Evol. Biol.* **3**: 139–44.

Charnov, E. L., 1991. Evolution of life history variation among female mammals. *Proc. Natl. Acad. Sci. USA* **88**: 1134–7.

Charnov, E. L., 1992. Allometric aspects of population dynamics: a symmetry approach. *Evol. Ecol.* **6**: 307–11.

Charnov, E. L. and Bull, J. J., 1977. When is sex environmentally determined? *Nature, Lond.* **266**: 828–30.

Charnov, E. L. and Bull, J. J., 1986. Sex allocation, pollinator attraction and fruit dispersal in cosexual plants. *J. Theor. Biol.* **118**: 321–5.

Charnov, E. L. and Bull, J. J., 1989*a*. Non-Fisherian sex ratios with sex change and environmental sex determination. *Nature, Lond.* **338**: 148–50

Charnov, E. L. and Bull, J. J., 1989*b*. The primary sex ratio under environmental sex determination. *J. Theor. Biol.* **139**: 431–6

Charnov, E. L. and Berrigan, D., 1990. Dimensionless numbers and life history evolution: age of maturity versus the adult lifespan. *Evol. Ecol.* **4**: 273–5.

Charnov, E. L. and Berrigan, D., 1991*a*. Dimensionless numbers and the assembly rules for life histories. *Phil. Trans. R. Soc. Lond. B* **332**: 41–8.

Charnov, E. L. and Berrigan, D., 1991*b*. Evolution of life history parameters in animals with indeterminate growth, particularly fish. *Evol. Ecol.* **5**: 63–8.

Charnov, E. L., Maynard Smith, J., and Bull, J. J., 1976. Why be an hermaphrodite? *Nature, Lond.* **263**: 125–6.

Charnov, E. L., Gotshall, D., and Robinson, J., 1978. Sex ratio: adaptive response to population fluctuation in Pandalid shrimp. *Science* **200**: 204–6.

Charnov, E. L., Los-Denhartogh, R. L., Jones, W. T., and Van den Assem., J., 1981. Sex ratio evolution in a variable environment. *Nature, Lond.* **289**: 27–33.

Choat, J. H. and Robertson, D. R., 1975. Protogynous hermaphroditism in fishes of the family Scaridae. In Reinboth, R. (ed.), *Intersexuality in the animal kingdom*, pp. 263–83. Springer-Verlag, Berlin.

Christie, J. R., 1929. Some observations of sex in the Mermithidae. *J. Exp. Zool.* **53**: 59–76.

Clark, A. B., 1978. Sex ratio and local resource competition in a prosiminian primate. *Science* **201**: 163–5.

Clutton-Brock. T. H., 1986. Sex ratio variation in birds. *Ibis* **128**: 317–29.

Clutton-Brock, T. H. (ed.), 1988. *Reproductive success: studies of individual variation in contrasting breeding systems*, University of Chicago Press.

Clutton-Brock, T., 1991. *The evolution of parental care*. Princeton University Press.

Clutton-Brock, T. H. and Albon, S. D., 1982. Parental investment in male and female offspring in mammals. In King's College Sociobiology Group (eds.), *Current problems in sociobiology*, pp. 223–47. Cambridge University Press.

Clutton-Brock, T. H. and Iason, G. R., 1986. Sex ratio variation in mammals. *Q. Rev. Biol.* **61**: 339–74.

Colby P. J. and Nepszy, S. J., 1981. Variation among stocks of walleye (*Stizostedion vitreum vitreum*): management implications. *Can. J. Fish. Aquat. Sci.* **38**: 1814.

Colby, P. J., McNicol, R. E., and Ryder, R. A., 1979. Synopsis of biological data on the walleye *Stizostedion v. vitreum* (Mitchell 1818). *FAO Fish Synopsis* **119**: 1–139.

Cole, L. C., 1954. The population consequences of life history phenomena. *Q. Rev. Biol.* **29**: 103–37.

Conover, D. O. and Heins, S. W., 1987. Adaptive variation in environmental and genetic sex determination in a fish. *Nature, Lond.* **326**: 496–8.

Cox, P. A., 1988. Hydrophilous pollination. *Annu. Rev. Ecol. Syst.* **19**: 261–80.

Cumaraswamy, A. and Bawa, K. S., 1989. Sex allocation and mating systems in pigeonpea, *Cajanus cajan* (Fabaceae). *Plant Syst. Evol.* **168**: 59–69.

Cushing, D. H., 1968. *Fisheries biology*. University of Wisconsin Press.

Damuth, J., 1987. Interspecific allometry of population density of mammals and other animals: the independence of body mass and population energy use. *Biol. J. Linn. Soc.* **31**: 193–246.

Darwin, C., 1871. *The descent of man, and selection in relation to sex*. Murray, London.

Deutsch, C. J., Haley, M. P., and LeBoeuf, B. J., 1990. Reproductive effort of male northern elephant seals: estimates from mass loss. *Can. J. Zool.* **68**: 2580–93.

Dunham, A. E., Miles, D. B., and Reznick, D. N., 1988. Life history patterns in squamate reptiles. In Gans, C. and Huey, R. B. (eds.), *Biology of the reptilia*, Vol. 16, pp. 441–552. Academic Press, New York.

Ebbert, M. A., 1993. Endosymbiotic sex ratio distorters in insects and mites. In Wrensch, D. L. and Ebbert, M. A. (eds.), *Evolution and diversity of sex ratio in insects and mites*, pp. 150–91. Chapman & Hall, London.

Ebert, T., 1975. Growth and mortality in post-larval echinoids. *Am. Zool.* **15**: 755–75.

Eisenberg, J. F., 1981. *The mammalian radiations*. University of Chicago Press.

Emlen, J. M., 1970. Age specificity in ecological theory. *Ecology* **51**: 588–601.

Farlow, J. O., 1976. A consideration of the trophic dynamics of a late cretaceous large-dinosaur community (Oldman Formation). *Ecology* **57**: 841–57.

Felsenstein, J., 1985. Phylogenies and the comparative method. *Am. Nat.* **125**: 1–15.

Fenchel, T., 1974. Intrinsic rate of natural increase: the relationship with body size. *Oecologia* **14**: 317–26.

Fischer, E. A., 1981. Sexual allocation in a simultaneously hermaphroditic coral reef fish. *Am. Nat.* **117**: 64–82.

Fischer, E. A., 1984. Local mate competition and sex allocation in simultaneous hermaphrodites. *Am. Nat.* **124**: 590–6.

Fisher, R. A., 1930. *The genetical theory of natural selection*. Oxford University Press. Reprinted by Dover Publications, New York, 1958.

Fleagle, J. G., 1985. Size and adaptation in primates. In Jungers, W. (ed.), *Size and scaling in primate biology*, pp. 1–20. Plenum Press, London.

Fowler, C. W., 1981. Density dependence as related to life history strategy. *Ecology* **62**: 602–10.

Fowler, C. W., 1987. A review of density dependence in populations of large mammals. In Genoways, H. (ed.), *Current mammalogy*, pp. 401–41. Plenum Press, New York.

Fowler, C. W., 1988. Population dynamics as related to rate of increase per generation. *Evol. Ecol.* **2**: 197–204.

Frank, S. A., 1983. A hierarchical view of sex-ratio patterns. *Fl. Entomol.* **66**: 42–75.

Frank, S. A., 1985. Hierarchical selection theory and sex ratios. II. On applying the theory, and a test with fig wasps. *Evolution* **39**: 949–64.

Frank, S. A., 1986*a*. Hierarchical selection theory and sex ratios. I. General solutions for structured populations. *Theor. Popul. Biol.* **29**: 312–42.

Frank, S. A., 1986*b*. The genetic value of sons and daughters. *Heredity* **56**: 351–4.

Frank, S. A., 1987. Individual and population sex allocation patterns. *Theor. Popul. Biol.* **31**: 47–74.

Frank, S. A., 1990. Sex allocation theory for birds and mammals. *Annu. Rev. Ecol. Syst.* **21**: 13–56.

Freeman, D. C., Harper, K. T., and Charnov, E. L., 1980. Sex change in plants: old and new observations and new hypotheses. *Oecologia* **47**: 222–32.

Fretwell, S. D. and Lucas, H. L., 1970. On territorial behavior and other factors influencing habitat distribution in birds. *Acta Biotheor.* **19**: 16–36.

Gaston, K. J., 1988. The intrinsic rates of increase of insects of different size. *Ecol. Entomol.* **14**: 399–409.

Ghiselin, M. T., 1969. The evolution of hermaphroditism among animals. *Quart. Rev. Biology* **44**: 189–208.

Godfray, H. C. J. and Parker, G. A., 1991. Clutch size, fecundity and parent–offspring conflict. *Phil. Trans. R. Soc. Lond. B* **332**: 67–79.

Goldman, D. A. and Willson, M. F., 1986. Sex allocation in functionally hermaphroditic plants: a review and critique. *Bot. Rev.* **52**: 157–94.

Gouyon, P.-H. and Couvet, D., 1987. A conflict between two sexes, females and hermaphrodites. In Stearns, S. (ed.), *The evolution of sex and its consequences*, pp. 243–62. Birkhauser Verlag, Basel.

Gowaty, P. A., 1991. Facultative manipulation of sex ratio in birds: rare or rarely observed? In Power, D. M. (ed.), *Current ornithology*, pp. 141–72. Plenum Press, New York.

Gross, M. R., 1979. Cuckoldry in sunfishes (*Lepomis*: Centrarchidae). *Can. J. Zool.* **57**: 1507–9.

Gross, M. R., 1982. Sneakers, satellites and parentals: polymorphic mating strategies in North American sunfishes. *Z. Tierpsychol.* **60**: 1–26.

Gross, M. R., 1984. Sunfish, salmon, and the evolution of alternative reproductive strategies and tactics in fishes. In Wooton, R. and Potts, G. (eds.), *Fish reproduction: strategies and tactics*, pp. 55–75. Academic Press, London.

Gross, M. R., 1985. Disruptive selection for alternative life histories in salmon. *Nature, Lond.* **313**: 47–8.

Gross, M. R., 1991a. Evolution of alternative reproductive strategies: Frequency dependent sexual selection in male bluegill sunfish. *Phil. Trans. R. Soc. Lond. B* **332**: 59–66.

Gross, M. R., 1991b. Salmon breeding behavior and life history evolution in changing environments. *Ecology* **72**: 1180–6.

Gross, M. R., 1993. Evolution of alternative reproductive tactics and strategies. Unpublished ms.

Gross, M. R. and Charnov, E. L., 1980. Alternative male life histories in bluegill sunfish. *Proc. Natl. Acad. Sci. USA* **77**: 6937–40.

Hamilton, W. D., 1964. The genetical evolution of social behavior. *J. Theor. Biol.* **7**: 1–52.

Hamilton, W. D., 1966. The moulding of senescence by natural selection. *J. Theor. Biol.* **12**: 12–45.

Hamilton, W. D., 1967. Extraordinary sex ratios. *Science* **156**: 477–88.

Hamilton, W. D., 1972. Altruism and related phenomena, mainly in social insects. *Annu. Rev. Ecol. Syst.* **3**: 193–232.

Hamilton, W. D., 1979. Wingless and fighting males in fig wasps and other insects. In Blum, M. S. and Blum, N. A. (eds.), *Sexual selection and reproductive competition in insects*, pp. 167–220. Academic Press, New York.

Harvey, P. H. and Clutton-Brock, T. H., 1985. Life history variation in primates. *Evolution* **39**: 559–81.

Harvey, P. H. and Nee, S., 1991. How to live like a mammal. *Nature, Lond.* **350**: 23–4.

Harvey, P. H. and Pagel, M. D., 1991. *The comparative method in evolutionary biology*. Oxford University Press.

Harvey, P. H. and Zammuto, R. M., 1985. Patterns of mortality and age at first reproduction in natural populations of mammals. *Nature, Lond.* **315**: 319–20.

Harvey, P. H., Read, A. F., and Promislow, D. E. L., 1989. Life history variation in placental mammals: unifying the data with theory. *Oxford Surv. Evol. Biol.* **6**: 13–31.

Harvey, P. H., Pagel, M. D., and Rees, J. A., 1991. Mammal metabolism and life histories. *Am. Nat.* **137**: 556–66.

Hennemann, W. W., 1983. Relationship among body mass, metabolic rate and the intrinsic rate of natural increase in mammals. *Oecologia* **56**: 104–8.

Heron, A. C., 1972. Population ecology of a colonizing species: the pelagic tunicate *Thalia democratica*. *Oecologia* **10**: 269–312.

Hoagland, K. E., 1978. Protandry and the evolution of environmentally-mediated sex change: a study of the Mollusca. *Malacologia* **17**: 365–91.

Hoenig, J. M., 1983. Empirical use of longevity data to estimate mortality rates. *Fish. Bull.* **82**: 898–903.

Hrdy, S. B., 1987. Sex-biased parental investment among primates and other mammals: a critical evaluation of the Triver–Willard hypothesis. In Gelles, R. and Lancaster, J. (eds.), *Child abuse and neglect: biosocial dimensions*, pp. 97–147. Aldine, New York.

Huger, A., Skinner, S. W., and Werren, J. H., 1985. Bacterial infections associated with the son-killer trait in the parasitoid wasp *Nasonia* (=*Mormoniella*) *vitripennis*. *J. Invertebr. Pathol.* **46**: 272–80.

Iwasa, Y., 1991. Sex change evolution and cost of reproduction. *Behav. Ecol.* **2**: 56–68.

Janzen, F. J. and Paukstis, G. L., 1991. Environmental sex determination in reptiles: ecology, evolution and experimental design. *Q. Rev. Biol.* **66**: 149–79.

Karlin, S. and Lessard, S., 1986. *Theoretical studies on sex ratio evolution.* Princeton University Press.

King, B. H., 1987. Offspring sex ratios in parasitoid wasps. *Q. Rev. Biol.* **62**: 367–96.

Kirkwood, T. B. L. and Rose, M. R., 1991. Evolution of senescence: late survival sacrificed for reproduction. *Phil. Trans. R. Soc. Lond. B* **332**: 15–24.

Kozlowski, J., 1992. Optimal allocation to growth and reproduction: implications for age and size at maturity. *Trends Ecol. Evol.* **7**: 15–19.

Kozlowski, J. and Wiegert, R. G., 1986. Optimal allocation of energy to growth and reproduction. *Theor. Popul. Biol.* **29**: 16–37.

Kozlowski, J. and Wiegert, R. G., 1987. Optimal age and size at maturity in annuals and perennials with determinate growth. *Evol. Ecol.* **1**: 231–44.

Krebs, J. R. and Davies, N. B. (eds.), 1987. *An introduction to behavioural ecology.* Blackwell, Oxford.

LaBarbera, M., 1989. Analyzing body size as a factor in ecology and evolution. *Annu. Rev. Ecol. Syst.* **20**: 97–117

Lack, D., 1954. *The natural regulation of animal numbers.* Oxford University Press.

Lande, R., 1982. A quantitative genetic theory of life history evolution. *Ecology* **63**: 607–15.

Lavigne, D. M., 1982. Similarity of energy budgets of animal populations. *J. Anim. Ecol.* **51**: 195–206.

Lee, P. C., Majluf, P., and Gordon, I. J., 1991. Growth, weaning and maternal investment from a comparative perspective. *J. Zool.* **225**: 99–114.

Leigh, E. G., Charnov, E. L., and Warner, R. R., 1976. Sex ratio, sex change and natural selection. *Proc. Natl. Acad. Sci. USA* **73**: 3656–60.

Lessells, C. M., 1991. The evolution of life histories. In Krebs, J. R. and Davies, N. B. (eds.), *Behavioural ecology*, pp. 32–65. Blackwell, Oxford.

Linstedt, S. L. and Calder, W. A., 1981. Body size, physiological time and longevity in homothermic animals. *Q. Rev. Biol.* **56**: 1–16.

Linstedt, S. L. and Swain, S. D., 1988. Body size as a constraint of design and function. In Boyce, M. (ed.), *Evolution of life histories of mammals*, pp. 93–106. Yale University Press.

Lloyd, D. G., 1984. Gender allocations in outcrossing cosexual plants. In Dirzo, R. and Sarukhán, J. (eds.), *Perspectives in plant population biology*, pp. 277–300. Sinauer, Sunderland, MA.

Lloyd, D. G., 1987. Allocations to pollen, seeds and pollination mechanisms in self-fertilizing plants. *Funct. Ecol.* **1**: 83–9.

Lloyd, D. G. and Bawa, K. S., 1984. Modification of the gender of seed plants in varying conditions. *Evol. Biol.* **17**: 255–338.

Loehle, C., 1988. Tree life histories: the role of defenses. *Can. J. For. Res.* **18**: 209–22.

Longhurst, A. R. and Pauly, D., 1987. *Ecology of tropical oceans.* Academic Press, New York.

MacArthur, R. H., 1965. Ecological consequences of natural selection. In Waterman, T. H. and Morowitz, H. (eds.), *Theoretical and mathematical biology*, pp. 388–97. Blaisdell, New York.

McMahon, T. A., 1973. Size and shape in biology. *Science* **179**: 1201–4.

Maekawa, K. and Hino, T., 1987. Effect of cannibalism on alternative life histories in charr. *Evolution* **41**: 1120–3.

Mangel, M. (ed.), 1990. *Sex allocation and sex change: experiments and models*. Lectures on mathematics in the life sciences, Vol. 22, American Mathematical Society, Providence, Rhode Island.

May, R. M., 1976. Estimating *r*: a pedagogical note. *Am. Nat.* **110**: 469–99.

Maynard Smith, J., 1982. *Evolution and the theory of games*. Cambridge University Press.

Medawar, P. B., 1952. *An unsolved problem in biology*. Lewis, London.

Metz, J., Nisbet, R., and Geritz, S., 1992. How should we define fitness for general ecological scenarios? *Trends Ecol. Evol.* **7**: 198–202.

Milinski, M. and Parker, G. A., 1991. Competition for resources. In Krebs, J. R. and Davies, N. B. (eds.), *Behavioral ecology*, pp. 137–68. Blackwell, Oxford.

Millar, J. S., 1977. Adaptive features of mammalian reproduction. *Evolution* **31**: 370–86.

Millar, J. S. and Zammuto, R. M., 1983. Life histories of mammals: an analysis of life tables. *Ecology* **64**: 631–5.

Moreau, J., Bambino, C., and Pauly, D., 1986. Indices of overall growth performance of 100 *Tilapia* (Cichlidae) populations. In Maclean, J. L., Dizon, L. B., and Hosillos, L. V. (eds.), *The first Asian fisheries forum*, pp. 201–6. Asian Fisheries Society, Manila.

Munro, J. L. and Pauly, D., 1983. A simple method for comparing the growth of fishes and invertebrates. *Fishbyte* **1**: 5–6.

Murray, B. G. and Nolan, V., 1989. The evolution of clutch size. I. An equation for predicting clutch size. *Evolution* **43**: 1699–1705.

Murray, B. G., Fitzpatrick, J. W., and Woolfenden, G. E., 1989. The evolution of clutch size. II. A test of the Murray–Nolan equation. *Evolution* **43**: 1706–11.

Nee, S., Mooers, A., and Harvey, P. H., 1992. Tempo and mode of evolution revealed from molecular phylogenies. *Proc. Natl. Acad. Sci. USA* **89**: 8322–6.

Nonacs, P., 1986. Ant reproductive strategies and sex allocation theory. *Q. Rev. Biol.* **61**: 1–21.

Orians, G. H., 1969. On the evolution of mating systems in birds and mammals. *Am. Nat.* **103**: 589–603.

Pagel, M. D. and Harvey, P. H., 1988. The taxon-level problem in the evolution of mammalian brain size: facts and artifacts. *Am. Nat.* **132**: 344–59.

Pais, A., 1986. *Inward bound*. Oxford University Press.

Parker, G. A., 1978. Searching for mates. In Krebs, J. R. and Davies, N. B. (eds.), *Behavioural ecology*, pp. 214–45. Blackwell, Oxford.

Parker, G. A., 1982. Phenotype-limited evolutionarily stable strategies. In King's College Sociobiology Group (eds.), *Current problems in sociobiology*, pp. 173–201. Cambridge University Press.

Parker, G. A., 1984*a*. Evolutionarily stable strategies. In Krebs, J. R. and Davies, N. B. (eds.), *Behavioural ecology: an evolutionary approach*, pp. 30–61. Sinauer, Sunderland, MA.

Parker, G. A., 1984*b*. The producer/scrounger model and its relevance to sexuality. In Barnard, C. J. (ed.), *Producers and scroungers: strategies for exploitation and parasitism*, pp. 127–53. Croom Helm, London.

Parker, G. A., 1992. The evolution of sexual size dimorphism in fish. *J. Fish Biol.*, **41** (Supplement): 1–20.

Parker, G. A., Baker, R. R., and Smith, V., 1972. The origin and evolution of gamete dimorphism and the male-female phenomen. *J. Theor. Biol.* **36**: 529–53.

Parker, W. S. and Plummer, M. V., 1987. Population ecology. In Seigel, R. A., Collins, J. T., and Novak, S. S. (eds.), *Snakes: ecology and evolutionary biology*, pp. 253–301. Macmillan, New York.

Partridge, L. and Sibly, R., 1991. Constraints in the evolution of life histories. *Phil. Trans. R. Soc. Lond. B* **332**: 3–13.

Pauly, D., 1978. A preliminary compilation of fish length growth parameters. *Ber. Inst. Meeresk. Univ. Kiel* **55**: 1–200.

Pauly, D., 1980. On the interrelationships between natural mortality, growth parameters, and mean environmental temperature in 175 fish stocks. *J. Cons. Cons. Int. Explor. Mer.* **39**: 175–92.

Pauly, D., 1981. The relationship between gill surface area and growth performance in fish: a generalization of von Bertalanffy's theory of growth. *Meeresforschung* **28**: 251–82.

Pauly, D. and Munro, J. L., 1984. Once more on the comparison of growth in fish and invertebrates. *Fishbyte* **2**: 21.

Peters, R. H., 1983. *The ecological implications of body size.* Cambridge University Press.

Peters, R. H., 1991. *A critique for ecology.* Cambridge University Press.

Petersen, C. W., 1990. Sex allocation in simultaneous hermaphrodites: testing local mate competition theory. In Mangel, M. (ed.), *Sex allocation and sex change: experiments and models*, pp. 183–205. Lectures on mathematics in the life sciences, Vol. 22, American Mathematical Society, Providence, Rhode Island.

Petersen, J. J., 1972. Factors affecting sex ratios of a mermithid parasite of mosquitoes. *J. Nematol.* **4**: 83–7.

Petersen, J. J., 1977. Effects of host size and parasite burden on sex ratio in the mosquitos *Octomymermis muspratti*. *J. Nematol.* **9**: 343–6.

Petersen, J. J., Chapman, H. C., and Woodward, D. B., 1968. The bionomics of a mermithid nematode of larval mosquitoes in southwestern Louisiana. *Mosq. News* **28**: 346–52.

Pianka, E. R., 1988. *Evolutionary ecology.* Harper & Row, New York.

Poinar, G. O., 1979. *Nematodes for biological control of insects.* CRC Press, Boca Raton, FL.

Policansky, D., 1982. Sex change in plants and animals. *Annu. Rev. Ecol. Syst.* **13**: 471–96.

Policansky, D. (ed.), 1987. Evolution, sex and sex allocation. *Bioscience* **37**: 466–506.

Promislow, D. E. L. and Harvey, P. H., 1990. Living fast and dying young: a comparative analysis of life history variation among mammals. *J. Zool.* **220**: 417–37.

Raimondi, P. T. and Martin, J. E., 1991. Evidence that mating group size affects allocation of reproductive resources in a simultaneous hermaphrodite. *Am. Nat.* **138**: 1206–17.

Ralston, S., 1986. Mortality rates of snapper and groupers. In Polovina, J. J. and Ralston, S. (eds.), *Proc. Workshop on the Biology of Tropical Groupers and Snappers, Honolulu, 1985*, pp. 16-24.

Read, A. F. and Harvey, P. H., 1989. Life history differences among the Eutherian radiations. *J. Zool.* **219**: 329-53.

Reiss, M. J., 1989. *The allometry of growth and reproduction*. Cambridge University Press.

Reiter, J. and LeBoeuf, B. J., 1991. Life history consequences of variation in age at primiparity in northern elephant seals. *Behav. Ecol. Sociobiol.* **28**: 153-60.

Reznick, D., 1985. Costs of reproduction: an evaluation of the empirical evidence. *Oikos* **44**: 257-67.

Ricker, W. E., 1973. Linear regressions in fishery research. *J. Fish. Res. Board Can.* **30**: 409-34.

Ricker, W. E., 1975. Computation and interpretation of biological statistics of fish populations. *Bull. Fish. Res. Board Can.* **191**: 1-382.

Ricklefs, R. E., 1969. Natural selection and the development of mortality rates in young birds. *Nature, Lond.* **223**: 922-5.

Ricklefs, R. E., 1977. On the evolution of reproductive strategies in birds: reproductive effort. *Am. Nat.* **111**: 453-78

Robertson, D. R. and Choat, J. H., 1973. Protogynous hermaphroditism and social systems in labrid fish. In Cameron, A. M. *et al.* (eds.), *Proc. 2nd Int. Symp. on Coral Reefs, Vol. 1*, pp. 217-25. Great Barrier Reef Committee, Brisbane.

Robertson, D. R. and Warner, R. R., 1978. Sexual patterns in the labroid fishes of the western Caribbean, II: The parrotfishes (Scaridae). *Smithson. Contrib. Zool.* **255**: 1-26.

Roff, D. A., 1980. A motion for the retirement of the von Bertalanffy function. *Can. J. Fish. Aquat. Sci.* **37**: 127-9.

Roff, D. A., 1982. Reproductive strategies in flatfish: a first synthesis. *Can. J. Fish. Aquat. Sci.* **39**: 1686-98.

Roff, D. A., 1983. An allocation model of growth and reproduction in fish. *Can. J. Fish. Aquat. Sci.* **40**: 1395-1404.

Roff, D. A., 1984. The evolution of life history parameters in teleosts. *Can. J. Fish. Aquat. Sci.* **41**: 989-1000.

Roff, D. A., 1986. Predicting body size with life history models. *Bioscience* **36**: 316-23.

Roff, D. A., 1992. *The evolution of life histories*. Chapman & Hall, London.

Rose, M. R., 1991. *Evolutionary biology of ageing*, Oxford University Press.

Ross, C., 1992. Environmental correlates of the intrinsic rate of natural increase in primates. *Oecologia* **90**: 383-90.

Rowe, D. K. and Thorpe, J. E., 1990. Differences in growth between maturing and non-maturing male Atlantic salmon, *Salmo salar*, parr. *J. Fish Biol.* **36**: 643-58.

Seber, G. A. F., 1973. *The estimation of animal abundance*. Griffin, London.

Seger, J., 1983. Partial bivoltinism may cause alternating sex-ratio biases that favour eusociality. *Nature, Lond.* **301**: 59-62.

Seger, J., 1991a. Conflict and cooperation in social insects. In Krebs, J. R. and Davies, N. B. (eds.), *Behavioural ecology: an evolutionary approach* (3rd edn.), pp. 338-73. Blackwell, Oxford.

Seger J., 1991*b*. Relatedness, sex ratios, and controls. *Science* **254**: 384.

Shapiro, D. Y., 1979. Social behavior, group structure and sex reversal in herma-phroditic fish. *Adv. Study Behav.* **10**: 43–102.

Shapiro, D. Y. and Lubbock, R., 1980. Group sex ratio and sex reversal. *J. Theor. Biol.* **82**: 411–26.

Shaw, R. F., 1958. The theoretical genetics of the sex ratio. *Genetics* **43**: 149–63.

Shaw, R. F. and Mohler, J. D., 1953. The selective advantage of the sex ratio. *Am. Nat.* **87**: 337–42.

Shine, R., 1978. Sexual size dimorphism and male combat in snakes. *Oecologia* **33**: 269–78.

Shine, R., 1979. Sexual selection and sexual dimorphism in the Amphibia. *Copeia* **1979**: 297–306.

Shine, R., 1980. 'Costs' of reproduction in reptiles. *Oecologia* **46**: 92–100.

Shine, R., 1988. The evolution of large body size in females: a critique of Darwin's 'fecundity advantage' model. *Am. Nat.* **131**: 124–31.

Shine, R., 1989. Ecological causes for the evolution of sexual dimorphism: a review of the evidence. *Q. Rev. Biol.* **64**: 419–61.

Shine, R., 1990. Proximate determinants of sexual differences in adult body size. *Am. Nat.* **135**: 278–83.

Shine, R. and Charnov, E. L., 1992. Patterns of survivorship, growth and maturation in snakes and lizards. *Am. Nat.* **139**: 1257–69.

Shine, R. and Schwarzkopf, L., 1992. The evolution of reproductive effort in lizards and snakes. *Evolution* **46**: 62–75.

Shumway, S. E., Perkins, H. C., Schick, D. F., and Stickney, A. P., 1985. Synopsis of biological data of the pink shrimp, *Pandalus borealis*. NOAA Tech. Rep. NMFS30. *FAO Fish Synopsis* **144**: 1–57

Sibly, R., Calow, P., and Nichols, N., 1985. Are patterns of growth adaptive? *J. Theor. Biol.* **112**: 553–74.

Sinclair, A. R. E., 1989. Population regulation in animals. In Cherrett, J. M. (ed.), *Ecological concepts*, pp. 197–241. Blackwell, Oxford.

Skinner, S. W., 1982. Maternally inherited sex ratio in the parasitoid wasp, *Nasonia vitripennis*. *Science* **215**: 1133–4.

Skinner, S. W., 1985. *Son-killer*: a third extrachromosomal factor affecting the sex ratio in the parasitoid wasp, *Nasonia* (=*Mormoniella*) *vitripennis*. *Genetics* **109**: 745–59.

Skinner, S. W., 1987. Paternal transmission of an extrachromosomal factor in a parasitoid wasp: Evolutionary implications. *Heredity* **59**: 47–53.

Slatkin, M., 1978. On the equilibration of fitness by natural selection. *Am. Nat.* **112**: 845–59.

Smith, F. E., 1954. Quantitative aspects of population growth. In Boel, E. (ed.), *Dynamics of growth processes*, pp. 277–94. Princeton University Press.

Southwood, T. R. E., 1981. Bionomic strategies and population parameters. In May, R. M. (ed.), *Theoretical ecology*, pp. 30–52. Sinauer, Sunderland, MA.

Stahl, W. R., 1962. Similarity and dimensional methods in biology. *Science* **137**: 205–12.

Stearns, S. C., 1992. *The evolution of life histories*, Oxford University Press.

Stearns, S. C. and Crandall, R. E., 1981. Quantitative predictions of delayed maturity. *Evolution* **35**: 455–63.

Stephens, D. W. and Dunbar, S. R., 1993. Dimensional analysis in behavioral ecology. *Behav. Ecol.*, **4**, 172–83.

Stewart, I. and Golubitsky, M., 1992. *Fearful symmetry*. Blackwell, Oxford.

Stubblefield, J. W., 1980. Theoretical elements of sex ratio evolution. Ph.D. Diss., Harvard University.

Sutherland, W. J., Grafen, A., and Harvey, P. H., 1986. Life history correlations and demography. *Nature, Lond.* **320**: 88.

Taylor, C. C., 1962. Growth equations with metabolic parameters. *J. Conserv.* **27**: 270–86.

Thompson, D. W., 1972. *On growth and form*, Vols 1 and 2. Cambridge University Press (originally published in 1917).

Trevelyan, R., Harvey, P. H., and Pagel, M. D., 1990. Metabolic rates and life histories in birds. *Funct. Ecol.* **4**: 135–41.

Trivers, R. L., 1972. Parental investment and sexual selection. In Campbell, B. (ed.), *Sexual selection and descent of man*. Aldine, Chicago, IL.

Trivers, R. L., 1974. Parent–offspring conflict. *Am. Zool.* **14**: 249–64.

Trivers, R. L. and Hare, H., 1976. Haplodiploidy and the evolution of the social insects. *Science* **191**: 249–63.

Trivers, R. L. and Willard, D. E., 1973. Natural selection of parental ability to vary the sex ratio of offspring. *Science* **179**: 90–2.

Vance, R. R., 1992. Optimal somatic growth and reproduction in a limited, constant environment: the general case. *J. Theor. Biol.* **157**: 51–70.

van den Assem, J., van Iersel, J. J. A., and Los-Denhartogh, R. L., 1989. Is being large more important for female than for male parasitic wasps? *Behavior* **108**: 160–95.

van Noordwijk, A. and De Jong, G., 1986. Acquisition and allocation of resources: their influence on variation in life history tactics. *Am. Nat.* **128**: 137–42.

Vollestad, L. A., L'Abee-Lund, J. H., and Saegrov, H., 1993. Dimensionless numbers and life history variation in Brown Trout: evaluation of a theory. *Evol. Ecol.*, **7**, 207–18.

Warner, R. R., 1975a. The adaptive significance of sequential hermaphroditism in animals. *Am. Nat.* **109**: 61–89.

Warner, R. R., 1975b. The reproductive biology of the protogynous hermaphrodite *Pimelometopon pulchrum* (Pisces: Labridae). *US Fish. Bull.* **73**: 262–83.

Warner, R. R., 1988a. Sex change and the size-advantage model. *Trends Ecol. Evol.* **3**: 133–6.

Warner, R. R., 1988b. Sex change in fishes: hypotheses, evidence and objections. *Environ. Biol. Fish.* **22**: 81–90.

Warner, R. R. and Robertson, D. R., 1978. Sexual patterns in the labroid fishes of the western Caribbean, I: The wrasses (Labridae). *Smithson. Contrib. Zool.* **254**: 1–27.

Warner, R. R., Robertson, D. R., and Leigh, E. G., Jr., 1975. Sex change and sexual selection. *Science* **190**: 633–8.

Werren, J. H., 1980. Sex ratio adaptions to local mate competition in a parasitic wasp. *Science* **208**: 1157–60.

Werren, J. H., 1983. Sex ratio evolution under local mate competition in a parasitic wasp. *Evolution* **37**: 116–24.

Werren, J. H., 1984a. Brood size and sex ratio regulation in the parasitic wasp *Nasonia vitripennis*. *Neth. J. Zool.* **34**: 151–74.

Werren, J. H., 1984b. A model for sex ratio selection in parasitic wasps: local mate competition and host quality effects. *Neth. J. Zool.* **34**(1): 81–96.

Werren, J. H., 1987a. Labile sex ratios in wasps and bees. *Bioscience* **37**: 498–506.

Werren, J. H., 1987b. The coevolution of autosomal and cytoplasmic sex ratio factors. *J. Theor. Biol.* **124**: 317–34.

Werren, J. H. and Charnov, E. L., 1978. Facultative sex ratio and population dynamics. *Nature, Lond.* **272**: 349–50.

Werren, J. H. and Simbolotti, P., 1989. Combined effects of host size and local mate competition on sex ratio evolution in *Lariophagus distinguendus*. *Evol. Ecol.* **3**: 203–14.

Werren, J. H. and van den Assem, J., 1986. Experimental analysis of a paternally inherited extrachromosomal factor. *Genetics* **114**: 217–33.

Werren, J. H., Skinner, S. W., and Charnov, E. L., 1981. Paternal inheritance of a daughterless sex ratio factor. *Nature, Lond.* **293**: 467–8.

Werren, J. H., Skinner, S. W., and Huger, A., 1986. Male-killing bacteria in a parasitic wasp. *Science* **231**: 990–2.

Werren, J. H., Nur, U., and Eickbush, D., 1987. An extrachromosomal factor which causes loss of paternal chromosomes. *Nature, Lond.* **327**: 75–6.

Werren, J. H., Nur, U., and Wu, C.-I., 1988. Selfish genetic elements. *Trends Ecol. Evol.* **3**: 297–302.

Western, D., 1979. Size, life history and ecology in mammals. *Afr. J. Ecol.* **17**: 185–204.

Western, D. and Ssemakula, J., 1982. Life history patterns in birds and mammals and their evolutionary interpretation. *Oecologia* **54**: 281–90.

Williams, G. C., 1957. Pleiotropy, natural selection and the evolution of senescence. *Evolution* **11**: 398–411.

Williams, G. C., 1966. Natural selection, the cost of reproduction and a refinement of Lack's principle. *Am. Nat.* **100**: 687–90.

Wooten, J. T., 1987. The effects of body mass, phylogeny, habitat and trophic level on mammalian age at first reproduction. *Evolution* **41**: 732–49.

Wrensch, D. L. and Ebbert, M. (eds.), 1993. *Evolution and diversity of sex ratio in insects and mites*. Chapman & Hall, London.

Author index

Subject index